CLEVER DOG!

LIFE LESSONS FROM THE WORLD'S MOST SUCCESSFUL ANIMAL

MASTER SALESMAN
MASTER SURVIVOR
MASTER TACTICIAN

MASTER NEGOTIATOR
MASTER LISTENER

MASTER OPPORTUNIST
MASTER OF INTEGRITY

RYAN O'MEARA

Hubble & Hattie

For more than nineteen years, the folk at Veloce have concentrated their publishing efforts on all-things automotive. Now, in a break with tradition, the company launches a new imprint for a new publishing genre!

The Hubble & Hattie imprint – so-called in memory of two, much-loved West Highland Terriers – will be the home of a range of books that cover all-things animal, produced to the same high quality of content and presentation as our motoring books, and offering the same great value for money.

More titles from Hubble & Hattie

This book is intended as a general reference guide. If you experience any sudden or serious health or behavioural concerns with your dog, you should consult a suitably qualified professional without delay.

WWW.HUBBLEANDHATTIE.COM

First published in March 2011 by Veloce Publishing Limited, Veloce House, Parkway Farm Business Park, Middle Farm Way, Poundbury, Dorchester, Dorset, DT1 3AR, England. Fax 01305 250479/e-mail info@veloce.co.uk/web www.veloce.co.uk or www.velocebooks.com.
ISBN: 978-1-845843-45-8 UPC: 6-36847-04345-2

Readers with ideas for books about animals, or animal-related topics, are invited to write to the editorial director of Veloce Publishing at the above address.
British Library Cataloguing in Publication Data – A catalogue record for this book is available from the British Library. Typesetting, design and page make-up all by Veloce Publishing Ltd on Apple Mac.
Printed in India by Replika Press

CONTENTS

ACKNOWLEDGEMENTS

This book is the result of a collaboration between dog enthusiasts, experts, owners and professionals from all four corners of the world, which perfectly highlights the immense presence of the canine as a universal figure of affection and joy.

With regard to the writing of this book I would like to extend my sincere thanks to my brother, Sean, for his valuable work in fact checking and editing; Anna Osbourne, for her significant contributions to some of the most fascinating elements of this project; Suzie Walker, for her fantastic contribution to the understanding of how dogs talk; and, last, but by no means least, I thank my wife and – even though they haven't yet mastered the art of reading – express my deep gratitude to all the dogs who provide us with a constant supply of truly glorious learning opportunities.

PREFACE

"Actions speak louder than words, and a smile says 'I like you. You make me happy. I am glad to see you.' This is why dogs are man's best friend. They are SO glad to see us that they almost jump out of their skin. So, when they are glad to see us, we are glad to see them. Don't you think it works in a human to human encounter? It does." – Dale Carnegie

It's actually in the dictionary, you know: 'man's best friend' – noun. The dog.

As one of the sixteen million people around the world to have a genuinely life-changing experience on the back of reading Dale Carnegie's classic self-help bible, *How to Win Friends and Influence People**, one passage in particular resonated with me so loudly I must have quoted it to just about everybody I know.

Why?

Well, not only because it's one hundred per cent true, and not only because it is one of those little nuggets of wisdom that can have a tangibly positive impact on one's personal outlook. No.

The reason I relay this particular passage from the world's most famous personal improvement book is because it bears out a belief I've had throughout my entire life ... dogs are gifted, at birth, with a dose of what I can only describe as pure magic. They do something to people that no other animal on the planet has ever been able to replicate in such consistently large numbers.

The passage I refer to is this:

"Why read this book to find out how to win friends? Why not study the technique of the greatest winner of friends the world has ever known? Who is he?

*How to Win Friends and Influence People. Dale Carnegie. © Simon & Schuster. First published 1937

4

"You may meet him tomorrow coming down the street. When you get within ten feet of him, he will begin to wag his tail. If you stop and pat him, he will almost jump out of his skin to show you how much he likes you.

"And you know that behind this show of affection on his part, there are no ulterior motives: he doesn't want to sell you any real estate, and he doesn't want to marry you. Did you ever stop to think that a dog is the only animal that doesn't have to work for a living?

"A hen has to lay eggs, a cow has to give milk, and a canary has to sing. But a dog makes his living by giving you nothing but love."

He was right, you know. The man who, quite literally, wrote the book on how to win friends, influence people and get ahead in life cited, as his best example, an animal who embodies the lessons Carnegie was teaching. For those who doubt the impact and influence of our friend the dog – and there are those who do – I ask this:

Why the dog? Of the millions of species of animals to inhabit this planet we share, how is it that one, and only one, inherited the title 'Man's best friend'?

We're now happy to simply accept the fact that the dog 'won,' and has his paws firmly planted under our table. But why? And how?

Surely, the animal most akin to us is an ape? It is, after all, only a few chromosomes away from BEING us! But we don't refer to the Chimp as 'Man's best friend,' and we don't have boarding kennels all over the world looking after people's pet Spider Monkeys. We don't have entire industries made up of behaviourists, groomers, minders, and the like, who are all set up to care for our millions of pet Macaques. Yet these are the animals who, theoretically, nay, even logically should have slotted in beside us as our animal companions of choice. They look like us, they even think a bit like us, they most surely remind us of ... us! But maybe it's for this very reason that the canine and not the ape has risen to the top table as our most favoured animal companion.

This book explores the journey of the dog, from wolf to man's best friend, inarguably the most prosperous animal, second only to man himself. More importantly, this book seeks to explain how our own species can learn and benefit from the dog's success secrets, as a resounding success he most definitely is and will continue to be.

Our friend the dog. Carnegie recognised his unique brand of magic. Millions and millions of people throughout the world have seen it, too, and our quest is to try and identify it, learn from it, and see how we can apply it to our own lives.

If my favourite passage from *How to Win Friend's and Influence People* has taught us anything, it's that we *can* and *will* win friends and influence people if we observe and learn from the master of this particular art.

He's a master salesman.
He's a master survivor.
He's a master tactician.
He's a master negotiator.
He's a master listener.
He's a master opportunist.
He's a master of integrity.

And yet, despite all of his mastery, he's more than happy to call US master.

Maybe, just maybe, that's our first lesson on how to succeed in life, the canine way ...

INTRODUCTION

The dog is undeniably the most successful domestic animal of all time. He shares his life with ours, has integrated into our society, and won the hearts and minds of millions of us.

We call ourselves dog lovers because we do sincerely love them: they are fully-fledged members of our family, and we have elevated them to positions of authority in the human world. Assistance dogs, protection dogs, detection dogs, companion dogs ... they all enhance our lives immeasurably.

In this book we will examine a whole raft of canine skills and talents, and try to piece together how and why the dog has enjoyed such immense success as a domesticated animal.

We will look at the dog as a problem solver, conflict resolver, and health asset. As a decision maker and hero. And as a loyal and trusted friend. We will look at the positive impact our friend, the canine, has had on human society generation after generation.

Throughout the creation of this work, we've had input from dog owners old and new; psychologists, dog behaviour experts, scientists, and many other observers with professional or personal interest in all-things canine.

We have followed the dog from his earliest known days right up to the present, charting his evolution, and even pointing out where he has been responsible for genuine revolution.

Clever Dog! is a compendium of life lessons we can learn from our dogs, based on a combination of what we know about them and – indeed – what we *think* we might know about them.

He's earned his epithet 'Man's best friend:' in this book, our aim is to examine how he did it and how we might emulate his talents in order to enjoy the same success in our lives.

Dedication
For Jackson, Chloe, Mia, and every other dog who has contributed to the most legendary human and animal relationship of all time

WHY PEOPLE LOVE DOGS

Understanding the legendary relationship between Man and Dog

"I think we are drawn to dogs because they are the uninhibited creatures we might be if we weren't certain we knew better." – George Bird Evans

Before we begin to examine some of the well-known reasons why people have grown to love dogs over the past ten to fifteen thousand years, we should first take a look at the dog's journey from wolf to Weimaraner.

Charting the rise and rise of the dog

Eavesdrop on any discussion about dogs and you will find discontent. The same can be said of the academics who have attempted to reach back in time to pluck the source of the canine out of the fossilised rocks in far away deserts. What came before the wolf? Can the Chihuahua and the St Bernard really share the same DNA? Do different breeds come from different wolves? Did we co-evolve, with man recognising and adapting to the benefits of his relationship with dogs?

The scientific community estimates that wolves and dogs began their divergence 15,000 years ago. Common ancestry from a single gene pool is still present at this time, but dogs and wolves effectively became different animals. Certain animals, such as ancestors of the Husky, German Shepherd, Malinois and Akita, remained close to the wolf in terms of geography and behaviour. These early dogs are known as proto dogs, and are the first generation of what we now know as the 'domestic' dog.

One of the main contributors to the domestic canine gene pool is the Indian wolf, direct descendants of which include many of today's wolves, plus the Dhole, Pariah dogs, and the Husky.

Ancestral species related to the European wolf bred with the now extinct Indian wolf (not to be confused with Himalayan wolves living in India) from which union the Saint Bernard, Pug, Bloodhound and Tibetan Mastiff can be traced. Although hugely different in size, that these dogs share a distinctive flat face is obvious; beginning life as one breed, they would gradually have diverged into the individual breeds they are today.

Meanwhile, in Europe, other European wolves were contributing to Spitz-type breeds. It's estimated that the Spitz was effectively a small wolf some 15,000 years ago, which diverged over time, producing smaller, whiter offspring. The first terriers and ancestors of herding dogs can also be traced to this wolf.

The crossbreeding of European and Chinese wolves is responsible for most toy and oriental breeds. These wolves were still in existence thousands of years after their descendants had become

CLEVER DOG!

established, and were able to mate with them to produce slightly larger, stronger versions of the dogs that were initially their descendants. The eastern Timber Wolf is a direct ancestor of dogs commonly known now as sled breeds.

Even as long ago as 12 thousand years, European and North American dogs were crossbreeding. It was possible for man to reach the New World from Eurasia by crossing the Bering Straight, a journey he made accompanied by dogs and wolves, as fossilised remains show.

Domestication

All over the world man was travelling with wolves and some proto-dogs. Natural selection played a large part in man's coming together with dogs. When early man learned to form settlements, wolves and dogs scavenged for food on the outskirts of camps, although only brave individuals would actually approach man and thus become noticed and initiate the symbiotic relationship that has continued to this day. The rest either remained wild or died out, but, in just two generations, dogs and wolves came to trust man completely.

Burial sites have been discovered globally that have contained the remains of dogs on their own, and dogs with humans. Many of these have been dated at around the time that humans experienced large population growth.

The burials appear to have been ceremonial, with dogs often positioned alongside humans, in the arms of humans, or at their feet. A 7000-year-old grave found in Sweden revealed the remains of a dog positioned at the feet of a man. The dog's neck had been broken, suggesting a ritualistic burial.

Other canine remains dating back as far as 8000 years suggest that injured dogs had been treated, aiding healing of broken bones: even euthanasia occurred during these times as dogs that were lame with arthritis and other illnesses had had their necks broken.

Since the domestication process began, man and dog have travelled together extensively; one reason why there are now so many breeds and breed types. Before dogs were split into working categories, they were categorised by the way they looked. Certain parts of the world had dogs that bore resemblance to the Timber Wolf, such as ancestors of the German Shepherd, whilst others deviated more quickly.

Man was quick to recognise the advantages offered by some physical traits inherited from the dog's ancestors, and it was from this point that deliberate crossbreeding became a part of the canine-human relationship. Prior to this, crossbreeding was a much more natural circumstance of evolution dictated by geography.

How dog breeds were created

'Breed' is most commonly used to describe a type of dog that displays characteristics interrelated to its name. But the concept of breed is far more complex than simply applying a label to a dog which looks a certain way.

As selective breeding processes became more widely practised by humans, certain characteristics were developed to aid them. Dogs were bred for three main purposes: to hunt, guard, and herd. As these functional breeds of dog became more established, crossbreeding was taking place in order to further refine certain necessary attributes.

Take early lupine domestication as an example. A large wolf which happened to be particularly quick over long distances, may be bred with a smaller example of the breed in an attempt to develop nimble, high stamina offspring for use when hunting. Overtly aggressive wolves would be bred to be as large as possible to produce effective guards for protection of human settlements. Naturally instinctive hunters would be bred to match the size of their quarry: selective breeding was in full swing.

This early form of selective breeding is how we eventually arrived at so many different breeds of dog. From the Yorkshire Terrier to the Bull Mastiff, selection by humans – as well as the occasional happy accident – has been the driving force behind the breeds we know and love today. As this process continued, physical and character traits of the originally domesticated wolves became more and more diluted, although, even today, every dog possesses some wolf-like characteristics.

Certain types have evolved to perform particular functions for mankind, and, although nowadays dogs are more commonly kept as companions, many breeds retain the original characteristics for which they were bred.

The hound group, for example, consists of two sections: sight hounds and scent hounds. The sight hound is one of the oldest types of dog still endemic today; quiet and graceful, he is bred to spot prey from a distance, then chase and despatch it. Sights hounds commonly kept as companions today include the Beagle, Greyhound, and Bassett Hound.

Terriers evolved in Britain from the hound group. Taking their name from the Latin 'to go to ground,' which is what they were bred to do, Terriers were originally used for hunting small quarry such as badgers, rats and rabbits. Like the hounds, terriers should have a strong work ethic, which, if not properly put to use, can occasionally result in unruly behaviour.

The toy group is the only one bred solely for the purpose of companionship. The Shih Tzu was favoured by Chinese emperors in the late nineteenth century as a companion in the palace, so was bred specifically for such purposes, as were other toy group members commonly referred to as 'lap dogs.'

The concept of what a breed actually is frequently causes confusion today, even amongst experts. What constitutes a separate breed if two dogs look the same and were bred for the same purpose? Why are Norwich and Norfolk Terriers classed as separate breeds now when before they were the same? Why do small cosmetic differences such as ear position result in a difference in breed, when major size difference does not? How can purebreeds resemble crossbreeds and crossbreeds be passed off as purebreeds?

All of these frequently-asked questions strengthen the argument that the concept of what a breed is is commonly misinterpreted.

A dog breed is probably best described as a grouping of descendants categorised using criteria relevant to its behavioural and physical qualities, desirable to those who refined the line of genetic descent.

Dogs have been making themselves 'useful' to humans for millions of years, as totems, gods, companions, guardians, hunters, hot water bottles, rescuers, guides, food, and the means of making a livelihood, amongst many other things.

But what is it about this particular species which has made it so appealing to us it has infiltrated almost every aspect of human life, and what can we, as humans, learn from this most adaptable of animals?

In the United Kingdom at Star Carr in Yorkshire, bones of dogs have been discovered in the tombs of humans dating back to 8500BC, along with tombs containing bones from eagles, songbirds and deer, suggesting that the animals were intended as some kind of totem or companion for the afterlife as well as helping these post Ice Age migrants hunt the giant auroch, elk, and wild boar. These settlers burned areas of scrub land to encourage plant shoots to tempt prey, and possibly domestic herds to eat, as well as to provide flatland on which to keep these predecessors of our own family companion, the dog. As Britain was at this point connected to mainland Europe, these settlers would have been travelling for thousands of miles to reach Yorkshire, all the while with the dogs in tow providing warmth and not just the means with which to hunt for food, but also, in an emergency, being the food themselves! Even today, expeditions with dog teams occasionally have to resort to culling some of the pack to avoid starvation.

This point in time appears to have been when man was most attuned to sharing his life with a canine: there was a common goal and purpose, shared reward, and mutual respect between the two species. There's no evidence of infantilism of the dog as surrogate child (pink fluffy dog sweatshirts have, as yet, not been unearthed ...), just a place within man's world, and a purpose to which the dog, with its relentless enthusiasm, capability and adaptability carried out with astonishing success and impact for humanity as a whole.

CLEVER DOG!

So what happened? Was it the hunters who first twigged that a wolf could help them hunt as it scavenged for bones at the edge of a settlement, attracted by its superior speed and graceful dispatch of prey?

Or was it a woman who took in (or was taken in?) by an appealing lost pup, who then grew to be her fierce protector? We'll never know for sure, but in my mind the original discourse went something like this:

Man: Wolf! Come and live with us and we shall provide you with the scraps of the animals you have hunted for us; we shall bury you alive with our dead as some kind of ritual, and wear your fur. Sound good?

Wolf: Not really. What's in it for me again?

Man: The protection and favour of the species about to take over the planet; we've got big plans you know. Plus, I won't kill you ... yet.

Wolf: Sigh.

So, despite an unpromising start to the relationship, the wolf managed to make himself so useful and appealing to humans that now one of his descendants (of the estimated 400 million) is probably snoozing comfortably at your feet as you read this ...

Our canine cousins

We sometimes forget today that our dog's canine cousins surround us. Let's examine the 'split' between domestic canine and its ever-so-close relative, the Red Wolf.

In a remote and relatively unpopulated region of north eastern North Carolina, approximately one hundred Red Wolves make their home; all that's left of a species which exists nowhere else in the world. The Red Wolf (Canis rufus), a smaller and more slender cousin of the better-known and more widely distributed Gray Wolf (Canis lupus), has the dubious distinction of being the first large carnivore to become extinct in the wild in North America. Thanks to US government restoration efforts, however, a small reintroduced population of these wolves currently eeks out a fragile existence in the marshy lowlands called Pocosin. Their near extinction and subsequent comeback represents a remarkable conservation success story.

Red Wolves once roamed throughout south eastern America, as far north as Pennsylvania and as far west as central Texas. Because of its wide distribution, the Red Wolf played a vital role in a variety of ecosystems, from swampy lowlands to forested mountains.

Shy and reclusive like its Gray Wolf cousin, the Red Wolf was nonetheless hunted, trapped, and poisoned until few remained. By the 1970s, only a remnant population of Red Wolves existed in marginal habitat along the Gulf Coast of south eastern Texas and south western Louisiana. In a recovery programme triggered by mandates of the Endangered Species Act, America's strongest wildlife conservation law, the last fourteen pure Red Wolves were captured by United States Fish and Wildlife Service biologists and placed in a captive breeding programme in a last-ditch attempt to save the species.

The Red Wolf's road from near extinction to recovery has been steady. However, the recovery effort has been plagued by biological and political hurdles, including scientific disputes, crossbreeding with Coyotes, political and legal opposition, and several ongoing threats to remaining habitat from urbanisation.

Repopulating the Red Wolf territories

Reintroduction to the wild began in the late 1980s with the successful but limited release of captive-born wolves on Bulls Island off South Carolina's Atlantic coast. This experiment was followed by the reintroduction of captive-born Red Wolves in Alligator River National Wildlife Refuge in 1987, and later into Pocosin Lakes National Wildlife Refuge, both in north eastern North Carolina. Notably, this was the first reintroduction of a species that was officially extinct in the wild. Red Wolves were also released into Great Smoky Mountains National Park, though, sadly, these animals could not find sufficient food or raise young successfully, and the restoration programme there was ended.

In the areas where Red Wolves are surviving, the ecosystem has benefited. By necessity, wolves tend to cull the weak, elderly and inferior animals from deer herds, thus making the remaining population larger and more robust. This is the way of nature: predators and prey co-evolving and co-existing so that the status quo is maintained. Area farmers have also found that Red Wolves help control Nutria, an invasive, non-native rodent species that damages crops and wetlands.

Thanks to the reintroduction programme, today, there are at least 100 Red Wolves residing in a five-county area of northeastern North Carolina. Population growth has been particularly strong in the past few years since officials began an intensive and successful management programme to prevent crossbreeding between Red Wolves and Coyotes. The US Fish and Wildlife Service recovery plan for the species calls for the release of Red Wolves at three separate sites to give a total wild population of 220 animals. Since the Great Smoky Mountains release did not succeed, officials will have to identify at least two additional sites in the southeast at which to release the wolves.

As successful as the reintroduction has been, the existing Red Wolf population is not immune to threat. Despite strong local opposition, the US navy recently proposed building a fighter-jet landing field right in the heart of Red Wolf country. On top of this, the navy has proposed designating special-use airspace over Pocosin Lakes and other areas where wolves reside. This action would give rise to low-level, high-speed flights by military jets, which, combined, would result in more than 30,000 flights a year in eastern North Carolina, bringing noise and air pollution and hindering essential access to the Red Wolves by government biologists. A coalition of conservation groups recently won a temporary court injunction against construction of the landing field, but the navy is appealing the decision ...

At the same time, Indiana-based Rose Acre Farms is proposing to build an egg factory in Red Wolf territory in North Carolina. The four million chickens factory would be one of the largest egg production operations ever built in the United States, and would have grave impact on air and water quality, and environmental integrity of the area. Increased road construction for the facility would fragment Red Wolf habitat, and the facility itself may attract wolves in search of prey, giving rise to interaction with humans in which the wolf would surely come off worse.

Hope

Despite the challenges, however, there *is* hope for the Red Wolf. A 2004 study by Defenders of Wildlife demonstrated that these creatures can be a huge economic boon to the rural areas they inhabit. The study surveyed visitors to North Carolina's popular Outer Banks beach resorts, less than an hour's drive from Red Wolf territory. Tourists were unanimous in their assertion that they would spend time and money to visit the natural areas that Red Wolves, Black Bears, Alligators, River Otters and other wildlife call home, thus generating thousands of dollars of income for the region's rural communities.

Conservationists are working to make these eco-tourism programmes a reality. The Red Wolf Coalition is moving forward with acquiring land and building a Red Wolf centre that would provide a home for a small pack of the animals which are unable to live in the wild, educate visitors about this unique natural resource, and draw tourism dollars to the region. In addition, conservationists are beginning to demand that the US government investigate other potential Red Wolf recovery sites from Pennsylvania to Florida. Because more reintroduction areas are essential for the long-term recovery of the Red Wolf, feasibility studies are warranted.

The Red Wolf is slowly regaining its rightful place in eastern North America's web of life. Strong public support for restoring wolves to the landscape has resulted in significant gains in the number of wolves, although obstacles to the continued recovery of Red Wolves remain. Conservation groups are working to overcome these challenges through ambitious outreach and education programmes, innovative solutions to on-the-ground conflicts, and partnerships with other stakeholder groups. Healthy, sustainable wolf populations not only bring ecological benefits, but economic rewards as well. To many people, restoring the wolf is not only a biological necessity, but a moral imperative. Because the species has suffered and been depleted at the hands of humans, many feel a sense of

CLEVER DOG!

obligation to future generations to restore what has been destroyed, and to preserve the wild lands on which great predators and wildlife depend. As we now know, the south eastern United States is not complete without its native canine inhabitants with which this area has been synonymous for generations.

Okay; why so much on the Red Wolf in a book all about dogs and how we can learn from them, you may ask? Well, as dog enthusiasts in their millions lavish time and money on their much loved pets, it's vital for us to recognise the inextricable link between our domestic friends and their (very) close relatives. The Red Wolf is a lesson in history; we nearly lost him. Yet our friend, the dog, is actually suffering from the opposite problem: there are too many of them.

As we seek to understand why the dog has achieved such popularity – to the point where his appeal has actually caused a population problem – we cannot possibly do so without discussing his close ancestors, but do this with a strong word of caution; dogs are NOT wolves. To put this into perspective, comparing a dog and the way he learns and behaves to a wolf is akin to trying to develop school lesson plans for children based on how youngsters learned and behaved several thousand years ago. In effect, whilst the genetic ties are close and indeed visible, we have to understand that, though the dog and the wolf share much, they are divided by thousands of years of evolution, and an accelerated 'design-by-humans' in the case of the selectively bred domestic canine.

It's vitally important that we understand and accept that not only is a dog no longer a wolf, in some cases we even have to re-train certain dog breeds to undertake their original function.

From wolf to Westie – the circle is complete

Elizabeth Everill is a graduate of Edinburgh University in Scotland. She spent much of her young adult life in the Middle East and Malaysia, and has owned and trained gun dogs for many years. Elizabeth recounts the amazing return to the wild of her own West Highland White Terrier, Jock, as we close the circle of how the wolf became the dog and the dog became a companion animal, which can still show his true, original colours as and when the opportunity arises.

Elizabeth recounts Jock's journey back to the West Highlands of Scotland:

"'I've taken a cottage in Achiltibuie, Wester Ross, for three weeks,' announced my younger son at the other end of the telephone. 'Thought you might like to join me and bring Jock. He would be returning to his place of origin, wouldn't he?'

"Yes he would, I mused to myself later, after accepting this welcome invitation. But how would Jock cope with such an environment? I could feel a project coming on ...

"Jock, a West Highland White Terrier, first came into my life as a puppy in 1997, when his owners were going on holiday. Over the subsequent years he came to me regularly, and a bond, stronger than any experienced with the working spaniels I had owned in the past, was formed. Jock, having been brought up as an urban dog, with nothing more asked of him than walking sedately to the local park and being a good house dog, had been trained well. This discipline stood him in good stead for living with us, but the rural life added another dimension – freedom to express his intelligence and his natural joie de vivre.

"I first took note of the West Highland White Terrier when one vied for top billing with the actor, Robert Carlyle, in the BBC's 1994 major drama series, Hamish Macbeth, set in the north west Highlands. Since then we've witnessed the deserved rise of Carlyle and the growing popularity throughout the world of this indomitable breed.

"In the authoritative book, Stonehenge and the Dog, published in 1887, there is no mention of the breed, only its ancestors, the Skye Terrier and the Scottish Terrier, both of which were kept as sporting dogs in the western Highlands of Scotland, unsurpassed in their tenacity when hunting rats, rabbits, foxes and badgers. Towards the end of the 19th century, Colonel E D Malcolm of Poltalloch, Argyll, worked Scottish and Cairn Terriers on his estate, and there gradually evolved a brown, beige or white dog he called the Poltalloch Terrier. Previously, on working estates, any white puppies in a litter were considered runts and despatched, but the story goes that some of the dark coloured dogs were shot as they worked the undergrowth in mistaken identity for vermin. So the white dogs became recognised as useful.

"In tandem with the evolution of the Poltalloch Terrier was the Roseneath Terrier. This dog's breeding came from the original Skye Terrier; not the one we see today in shows, but the one worked in ancient times on the remote Western Isles. The Roseneath had evolved as white in colour and was considered an admirable pet.

"In the book, *The West Highland White Terrier* by D Mary Dennis and Catherine Owen, research found that ships of the Spanish Armada, wrecked off the Western Isles, had carried small, white dogs to catch rats. It was, perhaps,the descendants of these found in Argyllshire in the 17th century that prompted the Scottish King, James I of England, to send a present of six little white 'earth' dogs to the King of France.

"Whatever the origin of the Westie, Mary Dennis suggests that, in naming it the West Highland White Terrier, a compromise was reached to include all the white dogs of the West Highlands. But certainly Colonel Malcolm of Poltalloch is credited with the development of the modern breed, leading to the first clubs in England and Scotland, formed in 1905.

"With only five months to go before our departure for Achiltibuie, I set about planning a training schedule for Jock. The terrain of north west Scotland is vastly different from that of Wiltshire where we live, and anyone who has spent a day walking over heather, bog and rock, will testify to the requisite fitness. Jock was no exception – and neither was I! Early mornings were devoted to a mile-long walk on a private road with no traffic. The surface hardened his paw pads and kept his toenails in trim. In the afternoons there were two or three mile hikes across Salisbury Plain, where the long tufted grass set challenges to Jock's muscles. We observed closely for any signs of fatigue, which could indicate a health problem. There never were; indeed, Jock seemed livelier on his return.

"Within weeks, Jock's little body was rock hard, but we maintained the regime. A month before our departure he had an MoT with the vet. Heart, ears, eyes, teeth, coat were checked, and he was pronounced fit.

"'One word of warning,' said the vet, quite intrigued by the project, 'the Highlands are thick with ticks and you already know that Jock has suffered infection from one of these in the past.'

"He parted Jock's hair at the neck and administered one drop of flea treatment from the phial. 'Repeat this just before you go,' he advised. 'Should any ticks get into the skin, they will die pretty quickly, but keep checking.'

"He was right. Two days after being amongst the heather, Jock had four ticks gorging on his blood, but they soon shrivelled and died.

"An hour before we set off to drive through the night, we gave Jock a prescribed tranquillizer and he slept for most of the 700 miles, but as he took so long to recover from the pill, we never repeated it. We took it in turns to drive, stopping every 3 of the 14 hours, to offer Jock water and the chance to relieve himself.

"Achduart Cottage has an idyllic setting amidst moor and mountain, with the sea of Loch Broom lapping 50 yards from the door. Jock was speedily launched into this environment, but he took to it cautiously, even after the effects of the tranquillizer had worn off. Negotiating the thick heather and bracken slowed him down, and the pools and bog puzzled him. Strangely enough, the rocky seashore proved no problem. Once he had picked up the scent of a family of Pine Martens which scavenged there, his inexperience was forgotten. The afternoons were tougher when we crossed the moors, but after ten days, Jock was bounding through the heather, wading pools and leaping from rocks with an aptitude which Colonel Malcolm would have been proud of.

"One day, on the lower slopes of Ben More Coigach, a pair of Golden Eagles appeared above us, wheeling and soaring. With his white coat standing out against the landscape, Jock was an easy target, and I decided to retreat with him, while my son continued to climb to the summit. Later, we found a sheep's body torn to shreds.

"The weather in Wester Ross can change in minutes and, although wet days were few, there were frequent gales bowling us over. Because of these we avoided precipitous areas, where

CLEVER DOG!

Jock would have stood no chance. Halfway through our stay, I introduced Jock to smaller hills, still steep and rocky, but not dangerous. He stuck close to me, sometimes moving a little way ahead, especially on the steepest parts, as if he were giving me encouragement, but mostly we climbed in tandem, which was thrilling.

"On Jock's first meeting with sheep, there was a stand-off, with Jock and the flock leader daring each other to move. Fearing future problems if the sheep backed down, I called Jock in and he came reluctantly, having lost face.

"Among some sand dunes we discovered huge burrows, at the entrance to which were fish bones – signifying otters. The early West Highland White Terrier was often used to flush out otters, which Jock promptly demonstrated by pushing his head well into the burrow. Otters, being a protected species, however, meant we had to interrupt Jock's fascination.

"Three weeks flew by. As I walked Jock down towards the shore on the last morning, he nosed his way confidently through the heather and bracken, leaping the burn in his stride. No longer urban, joyfully, he had acquired the ancient skills from, perhaps, some far distant dream."

Elizabeth's account teaches us that our relationship with man's best friend is largely one founded on human convenience. The dog is still quite capable of returning to its roots, and to really appreciate why we've adopted him with such devotion, we need only look to the fact that he is the ultimate model of adaptability.

He'll go anywhere with us, do pretty much anything for us, and can demonstrate just how easily he can switch from a return to his origins, to playing the role of a modern-day psychiatrist, doing little more than listen to our thoughts, lending a (seemingly) sympathetic ear, and lounging around with us on a lazy Sunday afternoon as we watch TV. He is, for want of a better explanation, whatever we want him to be at any one time. How can we not respect and be impressed by such a diverse set of attributes?

Our dogs convince us that being in our company leaves him delirious with delight, yet – inside that little body – is an animal that's really not too far removed from his wild and semi-wild cousins. The sketchy nature of our initial domestication of the dog begs the question: did we choose him – or did he choose us ...?

Lessons we can learn from dogs

The dog. An animal beloved by millions of people has undertaken a journey that has seen him evolve into an iconic, pop culture figure, and the deserving holder of the title 'man's best friend.' He's become a trusted member of our extended family, and we will go – in some cases literally – to the ends of the earth for him. He's a real friend who engenders friendship throughout the world, friendships that are life-long – how does he do that?

The three traits, psychological experts believe, that people want most from their relationships are –

• Empathy
• Non-possessive love
• Genuineness

– and the dog is almost a walking, barking, tail-wagging embodiment of these. Or, to be more accurate, we believe he is.

If we are to learn the real secrets of the canine success story and apply them to our own lives, our ability to acquire these three characteristics will determine how successful we are at emulating the dog's skill at gaining friends, winning affection, and having many people shower us with love. Empathy. Non-possessive love. Genuineness. These could very well be the three core attributes which meant that the dog succeeded where countless other species didn't in becoming firm fixtures in our lives and our affections.

Dogs have become part of our lives and our culture because they made themselves invaluable to us. By using their many abiities – hunter, guardian, companion, vermin catcher – they have helped ensure our success as a species. This approach can be applied in the workplace, too. By examining our performance and identifying those areas where we can do something that nobody else can, we are able to find ways of becoming invaluable. Don't be afraid to look outside of your existing job role and consider the immediate and long-term goals of your company. What does your company want to 'hunt,' who, or what is its quarry? New customers, more distributors, additional investment? Find a way to help it achieve these objectives. What does it need to protect itself from? Competitors taking clients, waste, litigation? How can you adapt to achieve these goals? What does the company want to avoid? Discrimination, distraction, boredom? Find ways to prevent these undesirable elements.

Since the industrial revolution, the dog as a worker has been pretty much redundant for a large percentage of the human race, but we've kept him on anyway, because he gives us something very special and unique: unconditional loyalty and love. People who offer the same to others usually have an abundance of friends and good family relationships.

We're very good at making unimportant things seem important. Sure, we've evolved so that survival and hunting aren't our primary urges any more, but we focus too heavily on small things and it can ruin our lives. Dogs happily carry on even after the most horrendous adversity (in our eyes, anyway). Taking a little bit of that attitude and approach and applying it to our own situation can help put things into perspective.

TWO

THE INFLUENTIAL DOG

How have dogs changed the world?

"If dogs talked, one of them would be president by now. Everybody likes dogs." – Dean Koontz

An interesting analysis of world leaders and power brokers shows that dog ownership is an incredibly common binding trait. Leaders, it would appear, love dogs. And when it comes to leaders, the greatest of them all, it could be argued, was Alexander the Great.

Alexander was only 32 years old when he died, but, during his short time on earth, he conquered most of the known world, and is exalted as one of history's most notable leaders. Alexander's conquests laid the foundations for modern civilisation, and brought together the continents of Europe and Asia. It says a deal about him that the revered Roman leader Julius Caesar – at the age of 32 – is said to have fallen to his knees and wept at a statue of Alexander, bemoaning his own lack of achievement in comparison (something he was to rectify in the coming years.)

However, without his beloved canine sidekick, Peritas, Alexander the Great might have been Alexander the Average. When the formidable leader was overwhelmed by the army of Persia's Darius III, Alexander's faithful companion is said to have attacked an elephant leading a charge against his master. Alexander subsequently survived the onslaught and lived to carry out his now immortal conquests. Western civilisation as we know it could have been immeasurably different, though, had not the dog risked his own life to save that of his illustrious master. So enamoured with his beloved companion was he that Alexander named a city after Peritas. As you can see, dogs have been making quite an impression on the human landscape for some considerable time, now.

Peritas was a Molossian – an ancient breed from which today's modern Mastiffs descend. The faithful dog followed Alexander in all his epic conquests, and risking his life in battle with his master was not a one-off occurrence.

When Alexander was trapped alone on the wrong side of the Mallians fortification, his men were blocked from reaching their leader. Leonnatus, one of Alexander's loyal officers who was fighting fiercely for his own survival during the battle, heard Peritas howl and bark from behind him, and, without looking over his shoulder, shouted "Go, Peritas! Run to Alexander!" The fearless hound ran through a great gathering of fighting men, and leapt into a huddle of Mallians who had just inflicted a serious javelin wound on Alexander. Peritas took down several of them, giving Alexander's troops the time they needed to reach their leader and prevent him from being killed.

Peritas, however, had not been as lucky; in saving his great master, he had been mortally wounded. With the last of his strength, he laid his head on the wounded king's lap and passed away, gazing into the eyes of his master.

If a dog has changed the course of history once, who's to say he won't do it again?

It's important, vitally important, we never forget what our dogs are prepared to do for us, including the ultimate sacrifice of laying down their lives for us. It's sadly ironic that some of today's breeds which share a link to the great, brave, loyal Molossers of the past are now demonised and derided as 'devil dogs' by the misinformed. Shockingly, there are even those who seek to kill dogs based on nothing more than their heritage.

In art, Alexander the Great is prominently depicted surrounded by dogs, for whom he evidently had great affection, perhaps recognising in them characteristics not usually associated with individuals who seek the presence of those with great power and influence. A dog shows loyalty not because it believes it will inherit great wealth and power by doing so, and his motives are largely transparent and free of conditions. Great leaders (and even bad ones) seem to share Alexander's empathy of canine companions.

Some of the most influential and powerful world leaders have turned to dogs for loyalty and companionship. Who knows what amazing secrets may have been whispered into the ear of a trusted canine confidante by some of history's most illustrious and notorious leaders?

To us, these individuals are historical icons. Some loved, some feared, and perhaps some misunderstood, but to the dogs that they shared their private moments with, they were merely a good friend, a master, even if one particular leader did deliberately poison his dog, and another cut off the ears of his. Let's take a look at some of the world's power brokers and their canine allies.

"To his dog, every man is Napoleon; hence the constant popularity of dogs." – Aldous Huxley, English novelist, 1894
This was undeniably true of Napoleon's dog, but the feeling may not have been reciprocated. Napoleon had a black and white dog called Sambo, whose ears he cut off, apparently for reasons of fashion, which gave his poor dog the look of a seal.

After Napoleon's death in May 1821, Sambo was taken to Europe by Countess Bertrand and her children, and now the stuffed animal resides in the Musée de l'Armée in Paris, ears still missing.

"If you want a friend in Washington, get a dog." – Harry Truman, 33rd President of the United States of America
Many of America's presidents have famously owned dogs, and, of course, the country's current leader, Barack Obama, is no exception. George W Bush was able to count two famous Scottish Terriers amongst his closest allies: Barney and Miss Beazley each had their own website on the official White House home page when their master was in residence, and Barney has many unofficial sites dedicated to him, too.

George W Bush's father, George Senior, had a Springer Spaniel called Millie who 'wrote' a book about her take on life as the first family's dog. Entitled *Millie's Book*, the account was ghostwritten by Barbara Bush and went on to be a bestseller.

"In a town in the east, the parishioners were visited upon, by a curious beast." – Justin Hawkins, vocalist in rock band, The Darkness
Black Shuck is a mythical dog whose master was a Viking by the name of Olaf the Fearless. Legend has it that Black Shuck haunts British shores, where he is rumoured to have been left behind by his Viking master in 787 AD. Those who catch sight of the spectral dog foresee the death of a loved one.

"The kids, like all kids, love the dog, and I just want to say this right now, that regardless of what they say about it, we're gonna keep it." – Richard Nixon, from the famous 'Checkers Speech'
Richard Nixon owes his political career to a little Cocker Spaniel named Checkers. At the 1952 Republican national convention, the then Senator, Richard M Nixon, was chosen to be the running mate of presidential candidate Dwight D Eisenhower.

Having been elected to Congress in 1946, Nixon quickly made a name for himself whilst

CLEVER DOG!

serving on the House Un-American Activities Committee, and, in 1950, was elected to the US Senate.

His rise to political prominence was nearly reversed forever in the midst of what became known as the Watergate Scandal, and Nixon used his dog in one of the most effective pieces of political spin in history.

With his wife sitting by his side, Nixon explained all of the details surrounding his finances, including the now-famous lines regarding his wife's "respectable Republican cloth coat" and the tale of his dog, named Checkers, who was given as a present to his young daughters. "I want to say right now that regardless of what they say, we're going to keep it."

From that point on, his words were immortalised as the 'Checkers Speech,' and Nixon went on to become the 37th President of America.

"I love animals, and especially dogs"– Adolf Hitler

Hitler has left an indelible thumbprint on world history for many grim reasons, but his love of animals is one of the less well-known aspects of his rather unique existence. Hitler had a sheepdog that he called Blondie, with whom he would spend a great deal of time, even during WWII.

Traudl Junge, Hitler's private secretary, explained how the feared Nazi dictator used to frolic with his dog. "Hitler's greatest pleasure was when Blondie would jump a few centimetres higher than the last time, and he would say that going out with his dog was the most relaxing thing he could do."

Blondie, who slept in Hitler's bedroom in his underground bunker, was to meet a grisly and untimely end at the hands of her own master. As the news reached Hitler that the Russians were approaching, he fed his dog a cyanide tablet and watched her die. Some say Hitler was testing the suicide pills on her first to make sure he would be dead before the Russian soldiers found him, rather than merely disabled, whilst others say he could not bear the thought of the soldiers tormenting or killing his dog, so he did it first.

Either way, poor Blondie, like so many others, died at the hands of one of the most hated people in history. It's often said that people come to resemble their dogs, and even adopt their companion's ways. In the case of the carnivorous, energetic, fun-loving sheepdog and her master, however, this probably couldn't be further from the truth.

White House dogs at a glance

Bill Clinton had a chocolate Labrador, which he called Buddy after an uncle who died shortly before he acquired his dog. Buddy's life was cut short when he was killed by a car, leading the former President to eulogise that he was a "loyal companion who would be truly missed."

Ronald Regan had two dogs during his time in the Oval Office: a Bouvier Des Flandres called Lucky and a King Charles Spaniel called Rex who replaced Lucky when she grew too big and was given to a friend.

Gerald Ford named his Golden Retriever Liberty. Liberty had a litter of pups whilst she lived at the White House.

Lyndon B Johnston caused a storm of controversy when he lifted his Beagle, called 'Him,' by the ears whilst on a press junket. The sister of 'Him,' named 'Her,' did not suffer the same fate.

Franklin D Roosevelt possibly had the most dogs at any one time whilst in office: his love of dogs was obvious. The lucky pups that got to live with the 32nd President were Major the German Shepherd, Meggie the Scottish Terrier, Winks the Llewellyn Setter, Tiny the English Sheepdog, President the Great Dane, and Fa la the Scottish Terrier.

Abraham Lincoln had a dog called Fido, whom he owned when he lived in Springfield, Illinois, before moving to the White House. As he was scared of the sound of bells, the President opted to leave Fido with his neighbour in Illinois, as he didn't want to see his dog traumatised. Fido died twelve months after his master's assassination.

"If you guys are going to be inhumane to my wife, you shouldn't pet my dog." – John Kennedy Jr
John Kennedy Jr, son of JFK, one of the most popular Presidents in American history, was never

without the company of dogs, thanks to his father's love of German Shepherds. JFK related in an interview with Larry King that his earliest memory was of a dog called Pushinka, who was given to his father by the then premier of the Soviet Union. Pushinka was the daughter of Laika, the first dog in space.

Royal pedigree

Dogs have featured strongly in the heritage of royal families. Certain breeds have taken their names from royals, whilst others wouldn't have existed at all had it not been by royal request.

To this day, all of the royal families – from the Plantagenets to the House of Orange – have had within their ranks dog fanciers of royal proportions. As we hear more and more today about so-called 'status dogs' (dogs which make a statement about the personality, character, or rank of his or her owner), it's easy to forget that dogs have always been used in this way, particularly by royals and royalists who established trends and made an impression on the public through their dogs of choice.

King Canute, the Great (1016-1035)

King Canute, the second Dane to rule England, was quite passionate about animal rights, a stance somewhat at odds with his passion for beheading any Londoner who still proclaimed Edmund II as King.

In being so passionate about the rights of rabbits and other small quarry, he gave rise to the concept of 'toy dogs' by passing a law stating that 'no dog that cannot fit through a gauge of eleven inches in diameter can be used for hunting.' This law was passed because the King saw no sport or fairness in large dogs such as Spaniels, hunting small quarry. In response, hunters began to breed only the smallest Spaniels, resulting in the breed known then as the Toy Spaniel.

Edward VI (1547-1543)

Edward VI was the long-awaited son of King Henry VIII. During his private education, the King was surrounded by 'privileged' children who were to share in his schooling. One of them, Barnaby Fitzpatrick, the cousin of the Earl of Ormonde, was appointed as Edward's whipping boy, meaning that any punishment due to the naughty young King (who ascended to the throne at age nine) was delivered instead to the hapless Barnaby, simply because it was not permissible to whip the King.

In spite of his rather unfortunate role, Barnaby and the King became good friends, and in later life Barnaby was given the title Baron of Upper Ossory. Another of the King's friends, Thomas Seymour, became jealous at apparently being ousted from the King's favours, and one night paid a visit to the King armed with a pistol. Unable to find the King, the unstable Seymour shot one of the King's beloved dogs instead. Fearing his punishment, he fled the school.

King James I (1603-1625)

King James, the ruling monarch of England, Scotland and Ireland during his reign, was a huge dog-lover. He had many hounds with whom he would go out hunting, and one in particular – Jewel – that he held in particular favour.

The Sovereign King spent much of his spare time with the hound. When she was found dead, killed by a single bullet, the King was outraged, until he discovered that the perpetrator of this crime was none other than his wife, Queen Anne, who had mistakenly shot Jewel whilst deer hunting. In an attempt to show his wife that he would love her nevertheless (and that her head would remain on her shoulders), he sent her a diamond as a tribute to the dead hound, and also told her she could have the London borough of Greenwich.

King Charles II (1660-1685)

King Charles II was made King after his father, Charles I, was executed following the English Civil War. Charles was famous for many things, including his magnificent ability to run parliament; known as the

CLEVER DOG!

Cavalier Parliament under his rule. He was also famous for his numerous illegitimate children, as well as his love of a particular breed of small dog.

The dog, a Toy Spaniel usually referred to simply as a 'comforte dog,' was often prescribed to the infirm by doctors as a means of treatment, and was renamed the Cavalier King Charles Spaniel in honour of the breed's most influential admirer.

Prince William of Orange (1689-1694)
This Dutch aristocrat, who went on to become King William III, was a famous admirer of Pugs. He took the dogs with him from England to Holland where he lived in the House of Orange. When the Dutch thwarted a Spanish invasion shortly before the Prince became King, he attributed this good fortune to the little dog, who went on to become the official symbol of the House of Orange.

Queen Charlotte (1760-1820), wife of King George III
Upon her marriage to King George III, Charlotte brought with her to the royal home some large, white, Spitz-type dogs. These dogs were pets to her, but her husband showed little interest in her hobby. It wasn't until a future monarch decided to 'improve' on the dogs by selective breeding that this particular dog became established as the Spitz breed we know today.

Queen Victoria (1837-1901)
Queen Victoria is responsible for one of the most popular toy breeds in existence today. After inheriting some of the royal dogs that were doted on by Queen Charllotte, wife of her predecessor to the throne, Victoria kept the dogs as pets. Upon the death of her husband, Prince Albert, in 1861, she turned much of her attention to tending to these dogs. Unhappy with their size, the Queen bred the dogs with smaller breeds to establish the breed we know today as the Pomeranian.

King Edward VII (1901-1910)
King Edward was quite the trendsetter amongst the British aristocracy. Noted for his foppish and flamboyant dress, and his many sexual indiscretions, the eldest son of Queen Victoria was somewhat of a rebel. Up to and during his reign, Edward was closely watched by his parents and their aides, and his behaviour was often criticised as being too risqué for a future King. Amongst his passions were gambling, sport, travel, women, and his Wire Fox Terrier called Caesar.

Caesar, too, managed to write himself into the history books, when, upon the death of King Edward in 1910, he was seen following the funeral procession, all by himself.

King George V (1910-1936)
King George V, grandfather to the current English Queen, founded the house of Windsor. A lover of Sealyham Terriers and Cairn Terriers, many royally commissioned paintings show him with his dogs.

Queen Elizabeth II (1952-present)
The reigning monarch is perhaps the most well-known dog lover of the present Royal Family. Elizabeth has owned over thirty Corgis, the first being Susan, who was given to her on her eighteenth birthday by her father, King George VI. The Corgis are, perhaps, the most well-known of the Queen's dogs. On Christmas Eve of 2003, one of the Queen's 'oldest and dearest' Corgis, Pharos, was attacked and killed by an English Bull Terrier owned by Princess Anne. The dog suspected of the attack, named Dotty, had courted controversy previously after landing the Princess in court following an attack on two youngsters in a park. It was later revealed that it wasn't Dotty who had killed Pharos, but a dog called Florence, who was said to have been startled upon arriving at Sandringham as the Queen and her army of Corgis had gone to greet Princess Anne.

Given the obvious attraction of dogs to some of the world's most influential people, it's little wonder that dogs have played such a pivotal role in moments of great importance throughout man's modern history.

Nowadays, we accept as normal the presence of many different types of working dog in society – service dogs, assistance dogs, detection dogs, and protection dogs. As our canine relationships have evolved, we've learned to elicit even more from the dog, and it's sobering to reflect on how differently history might have evolved had it not been for man's relationship with him.

Would Alexander the Great have lived long enough to leave his indelible mark on the world? Would the outcome of major battles and wars have been different? We have considered how the dog's appeal spans all social and financial standing; he is as important to people of great power, wealth and influence as he is to those who may not even have a home to call their own. And, regardless of whether you're a world leader or a homeless transient, a passion and interest in dogs provides a common bond. We must count ourselves most fortunate that the loyalty of a dog does not depend on wealth, social standing or world influence: to every dog, his master is king.

Why do leaders love dogs?

Most people have a dog because, as a companion, he offers much that we value: unconditional love, loyalty, trust, friendship, and much, much more. Some people maybe had a dog when growing up, and discovered that life without a dog is not as much fun as life with one; some always wanted a dog of their own, so, as soon as they left home, they got one. Others turn to a dog when it feels that there is something missing in their lives.

Not surprisingly, our world leaders have exactly the same reasons and motives for wanting canine companionship as the rest of the population, and it could be that their need for that companionship is more keenly felt. The life of a leader – be he or she head of a large company, lead singer in a pop band, or ruler of a country – can, at times, be isolated and lonely.

The effort to which people will go to include a dog in their hectic, globetrotting lives suggests that there is a need so acute that they are somehow incomplete until it is realised.

Humphrey Bogart, named the greatest male star of all time by the American Film Institute, probably epitomised the sentiments of all dog owners, in his own inimitable way, when neighbours complained about his dog's barking.

"What son of a bitch doesn't like dogs? What sort of monster is he? He ought to be glad to hear the sound of dogs barking."

One-time owner of five dogs, Bogart had a largely unhappy marriage to his third wife, Maya Methot, and when going through the divorce, was said to have fought tooth and nail to keep all the dogs with him, in spite of his busy lifestyle. He succeeded and went on to marry Lauren Bacall, with whom he acquired another two dogs. Throughout his professional life, as wives came and went, Bogart's dogs always stayed with him.

Elvis Presley led a generation to rebel and cut loose during a time where his nation was at war, and his homeland was peppered with civil unrest. Dogs were a huge part of Presley's often troubled, high profile life, as he sang about them, performed with them, and confided in them. Elvis never had a dog as a child, though had always wanted one. By the time he eventually got a small mongrel called Sweet Pea, Elvis was a confirmed dog lover, who went on to have Great Danes, Poodles, and in his final, uneasy years, a Chow, who accompanied him everywhere, even on stage.

JFK was allergic to dogs, but this didn't stop him keeping them as companions. When he wanted to be alone with his wife to talk privately, he would simply ask her to accompany him as he walked his two German Shepherds, Charlie and Clipper. The things those dogs must have been privy to ...

Nobody could begin their meal until Rufus, Poodle companion of Winston Churchill, had been served. Famed among his peers for his calming effect on all animals, Winston apparently regarded Rufus more as a confidante than a pet. Not shy of conducting one-sided conversations with Rufus, even in front of members of parliament, he famously covered the dog's eyes as they were watching a scene in a film where a dog died. Churchill whispered to Rufus, "Don't look now, dear, I'll tell you about it afterwards."

During the German occupation of Paris in WWII, a stroll without an obvious purpose was

CLEVER DOG!

viewed with suspicion, but for Pablo Picasso, a walk alongside the Seine was justified by the presence of his first dog, an Afghan Hound called Kazbek. Picasso and Kazbek sat through the war together; he told friends he needed to take in the beauty of the city and owed a debt of gratitude to Kazbek for allowing him to do so. Picasso owned many more dogs after Kazbek, including an incontinent Dalmatian which appeared in some of his later works.

When the Monica Lewinsky scandal broke across the USA, Bill Clinton's wife Hillary commented "The only member of the Clinton family to keep the President company was Buddy," their chocolate Labrador. Buddy was by Clinton's side as he bombed Iraq, acted as alibi in the infidelity allegation, and there when he stepped down as President. On Buddy's death, Clinton described him as "a loyal companion who brought much joy."

"Affection without ambivalence" is what the father of psychoanalysis, Sigmund Freud, said he got from his loyal Chow Jo Fi. Freud valued his Chow's 'one man dog' attitude and his unfailing loyalty. Jo Fi became synonymous with Freud, and even used to sit in on his therapy sessions as Freud believed his dog had the ability to calm distressed patients.

Former President Franklin D Roosevelt, upon joining with the allied forces after the Japanese attack on Pearl Harbour, was so concerned for the safety of his Scottish Terrier, Fala, that he had a horoscope drawn up for his dog by a local astrologer. In 1941, this was not a normal thing for an American President to be doing. There is now a statue of Fala, who outlived his master, at the FDR memorial in Washington.

Cultural leaders often take inspiration from their dogs. Jerry Leiber and Mike Stoller did so with *Hound Dog*, sung so memorably by Elvis, and so did legendary designer Yves Saint Laurent. Verging on the obsessed about French Bulldogs, the Frenchman saw his association with the dogs immortalised by Andy Warhol's series of portraits. The designer took his dogs everywhere, and upset models when the animals drooled, not on Saint Lauren's pieces, but the models' own clothes.

Another cultural and business leader who saw his dog immortalised in pop culture is Richard Branson, at one point, synonymous with his late Irish Wolfhound, Bootleg. Bootleg lived at The Manor, where artists signed to Branson's Virgin Records would live and record. After complaints from neighbours about the late night Black Sabbath recording sessions, the band seemed to think that it wasn't its music that was disturbing the neighbours, but the dog barking in response to its loud riffage. The band wrote a song in Bootleg's honour called *Digital Bitch*.

Regardless of popularity, success or profile, this selection of little-known tales about leaders and their dogs shows just how much a dog can be relied on by those in the public eye. Keep in mind, therefore, that behind the next news story we hear there may be an untold tale of a dog sitting stoically by the side of an unpopular, lambasted or hunted man or woman, just one reason why leaders need their dogs.

The Dogs of War

From Biblical times to the modern day, the history of the world has been shaped by war. From the Roman conquests to modern and sophisticated warfare, the battles have been bloody and gruesome, but all of them have put man side-by-side with his dog. Since the forming of the Persian empire in 550 BC, there hasn't been a war where dogs haven't served with their masters.

In November of 2004, a monument was unveiled by Princess Anne in Park Lane to commemorate the animals that served during many of the world's wars. The first of its kind, it includes a bronze rendering of a dog, representing all dogs to have received the Dicken Medal for bravery, and those who have served the military.

The Dicken Medal is awarded by the People's Dispensary for Sick Animals, and is the equivalent of the Victoria Cross for bravery. A recent recipient of this award is Buster. A six-year-old Springer Spaniel serving in Iraq, Buster made a very important discovery in an apparently disused house in Safwan, southern Iraq, when he located a cache of contraband that was to enable his handler and owner, Sgt Danny Morgan of the Royal Army Veterinary Corps, to gather intelligence that would lead to the arrest of sixteen of Saddam Hussein's supporters.

Buster's story and many like his go relatively untold when compared to those of soldiers and civilians, but they are testament to the loyalty of the dogs who unhesitatingly lay down their lives for their masters. Thrown into unfamiliar and frightening surroundings, those dogs who have been to war and not returned are referred to by sentimental terms such as 'heroes defending their country' and 'brave soldiers fighting for their people,' but they are not. They are dogs doing what dogs do; honouring the friendship between themselves and their masters to the very end. This is not to diminish the glory that the dogs earn but to credit it not to the turmoils and emotion of war, but the simple criteria by which all dogs live: to protect their master.

With this in mind it's all the more impressive to hear of brave dogs that carry on their roles as bomb detector, guardian, medical supplies carrier, and intelligence gatherer whilst injured or separate to their master.

During the war in Vietnam, the American government was criticized by the army's dog handlers for treating the dogs as 'equipment.' Dogs would be assigned to handlers during military service, and many would be sent out to serve. Stories of soldiers being forced to leave their dogs behind in the war zones to continue service until either the war ended or they were killed caused severe distress to certain American service men. Some even refused to return to their families unless they could take their dogs, in many cases dogs to whom they owed their lives.

Although dogs have been influential in war, war has had an effect on the development of dogs. Germans Shepards are often referred to as 'Alsatians' (the breed originated in Alsace). Although never officially known by this name, Americans adopted the moniker for political reasons during the two world wars. During times of conflict, Americans would always try to disassociate themselves from anything that had even the most tenuous link to the enemy, such as changing frankfurters to hotdogs and French fries to freedom fries.

Dogs in war zones are used not only to protect their handler, but also assist in protecting civilians. Major Richard Pope was serving in Kosovo with the Royal Army Veterinary Corps as part of the Commanding Multi National Working Dog Support Unit in 2002, when the skills of his four-and-half-year-old German Shepherd, Charlie, proved vital to maintaining peace.

Says Richard: "I had been working out in northern Kosovo on a peace maintenance mission, and received a call very late one night telling me I needed to track 'persons unknown' back to the Serbian border. An Albanian farmer had been receiving intimidating threats from two men who were believed to have come over the border from Serbia. It was my job to establish whether there was anyone coming from over the border. The farmer showed me the area where he believed the men had last been, and Charlie picked up their scent and tracked it back all the way to the border, following their exact movements. I had total trust in Charlie's actions, I did not know whether we would come across these men and would be shot at, but we made sure that the scent went from the farm all the way up to the border. This enabled us to report an infringement to the UN which was essential to the peace-keeping mission."

During more recent conflicts, dogs have had significant roles in preventing deaths and casualties through explosive detection. In 1971, British forces trained bomb detection dogs for service in Northern Ireland. Since then, the Bureau for Alcohol, Tobacco and Firearms has begun an anti-terrorism dog training campaign geared toward reducing terrorist activities through actively identifying explosives.

Dogs have been by our side as we've fought, planned, developed, evolved and made incredible scientific breakthroughs. They have no political or military alliances over and above wanting to please us, yet it's a fact that the world we know today would be very different had it not been for some pivotal moments in history when dogs were either in the thick of the action, or patiently listening to the musings and ruminations of their masters.

Lessons we can learn from dogs

Influence and power are subjective. In the eyes of a true friend, loyalty is not generated by virtue of what material gains or status can be bestowed in return for favours. Regardless of one's standing

CLEVER DOG!

in the world, having a loyal, close confidante to share life's problems with has a value that exceeds money. Dogs have managed to change the world not by setting out to become heroes, but simply by being themselves, demonstrating by example their natural instinct to protect with loyalty, devotion and selflessness.

Most of us have a 'leader' in our work lives; be it the boss or a mentor, there's usually somebody we strive to please. Many people make the mistake of trying to emulate those that they need to please, but why would your boss need another person exactly like him or herself? Finding out what they lack is more important, and filling that gap by offering something that they aren't getting. Could it be honesty? Are they surrounded by 'yes' men? Your dog would certainly let you know if you were late with his dinner, whether you're Prime Minister, head of a Fortune 500 company, or a humble farmer, he'd let you know. Does your leader need someone who simply listens? Think about what you value most from your dog and offer that.

Visit Hubble and Hattie on the web: www.hubbleandhattie.com and
www.hubbleandhattie.blogspot.com
Details of all books • New book news • Special offers • Gift vouchers

THREE

LEADERSHIP:
THE CANINE WAY
Understanding canine social structures and hierarchies

"The dog is an animal which seems to fundamentally understand that we're not here for a long time, we're here for a good time!" – Anonymous

Dogs use sophisticated forms of social cognition and communication, exhibiting various postures and different methods of non-verbal conversation. These abilities make them trainable, playful and adaptable to human households and social situations, as well as enable them to enjoy a unique relationship with humans.

There's a common misconception that man domesticated the wolf, which then became a dog; in fact, this evolution coincided with the emergence of the town dump, which presented opportunistic scavengers, such as wolves, with a new source of easily accessible food. Wolves soon learned that if they could overcome their innate fear of humans they would have the advantage of being able to eat more. It's amazing to think that, were it not for these communal refuse areas, the most famous animal/human relationship on the planet may never have actually come about ...

In ancient times, anthropologists concluded that the relationship between humans and dogs was mutually beneficial. Domesticated wolves and descendant dogs benefited from the humans' use of tools which enabled them to build shelter and make weapons; in turn, dogs assisted in hunting by using their keen sense of hearing and smell, as well as providing warmth, and protection. Anthropologists Paul Tacon and Colin Pardoe stated that the effect of human-canine cohabitation on humans has been profound, possibly giving rise to a shift to large game hunting, the establishment of territories, and negotiating partnership bonds with dogs in order to live together harmoniously.

From time immemorial, humans and animals have tended to live and travel in groups; in herds, colonies, and bands. For dogs, the basic social unit is the pack: tight-knit units consisting of individuals ranked in a linear hierarchy. Many studies show that domestic dogs adhere to a hierarchy centred on an 'Alpha-Beta-Omega' structure. While wolves and dogs may have similar genetic traits, studies indicate that there are distinct differences in their behaviour.

A study on dogs by Dr Frank Beach at Yale and UC Berkeley noted that male dogs have a more rigid hierarchy compared to females. In a dog pack, puppies have a licence to do whatever they want, at least until they reach four months, when older, 'middle-rank' dogs will harass and torture the youngsters to ensure that each takes its rightful place in the hierarchy – at the bottom! The study found that 'middle-rank' canines tend to fight among themselves as their status is insecure. Low-rank animals, such as a young dog, do not usually fight with middle-rank pack members; they know their place and would not normally do anything radical to advance their position.

CLEVER DOG!

Beach noted, however, that there is no physical domination among dogs. Most dogs who are regarded as leaders of the pack, or 'alpha dogs,' tend to be benevolent principals, confident in their position. In the case of dogs, the term 'alpha' does not usually apply to animals that are physically dominant, but rather those that control valued resources, such as food, and who have the right to mate with the females in the pack.

Other studies suggest that male and female dogs will compete for resources such as toys and food, albeit in a non-violent manner. Minor tussles among dogs will usually be about access to or control of valued resources. If you have more than one dog, the one that has not been neutered or spayed will serve as the dominant dog. In cases where both dogs are intact, the male will usually be the dominant dog. Dogs that have been neutered or spayed will usually be subordinate to an intact male or female canine. Older dogs, meanwhile, are regarded as less dominant than younger dogs in their prime. Citing studies, the Association of Pet Dog Trainers (ADPT) notes that dogs which use aggression to get their way are not necessarily displaying dominance; such behaviour could be the result of anxiety.

Unlike their domesticated dog cousins, wolves have an intense, intuitive drive which ensures that the pack is in order all the time. According to scientist L David Mech, wild wolf packs can comprise anything from two to twelve members; keeping the pack in order is crucial as the wolves depend on each other to survive in the wild. Mech found that the natural wolf pack is centred around a family, which consists of a breeding pair of adult wolves and their litter. Mech's study indicated that the 'alpha' wolves are the parent wolves. The formation and structure of a pack resembles that of our own families. While social hierarchies exist, Mech claims these are not related to aggression.

There is a more practical aspect to the pack, which is that the alpha male and alpha female will make decisions for the entire pack. And, of course, wolves have to work together in order to hunt and bring down big animals such as Caribou, which will provide meat for all pack members.

Within a pack, common alpha behaviour would include licking up (wherein the wolf begs for food); pinning, and standing over another wolf to mark territory. Passive signals would include the animal adopting a posture physically lower than other pack members, such as crouching or rolling over on their back and exposing the vulnerable abdomen, lowering the tail and flattening the ears.

Early dog training techniques were based largely on the pack principle. Companion dogs are generally considered to be a member of the family or a working partner. When properly trained and educated, dogs will take their cue from you, as alpha, and will learn how to behave appropriately inside and outside of the home. The problems can start when you humanise your four-legged friend too much, allowing him to eat from the table, or making a mess of your favourite chair. These liberties could give him or her the impression that they can dominate you.

Not always, but in some cases your dog could try to undermine your leadership, behaving like a bratty child in an effort to challenge you. And you know you're really in trouble when your dog growls at anyone (including you) who approaches his food dish, or touches one of his toys. The next stage on could be that he bites someone ...

Correcting behavioural problems using confrontational training techniques, according to the Association of Pet Dog Trainers, could instil fear in your dog, and worse, cause him to react aggressively. A frightened dog will protect himself by using his teeth. Grabbing a dog and forcing him down, or gripping a puppy's muzzle could elicit a 'flight-or-fight' response: the animal will either freeze in fear, try to run away, or fight to save his skin. An 'alpha roll' (flipping a dog onto his back and holding him in that position, sometimes by the throat), or a scruff shake (grasping the scruff of the dog's neck between his ears and above his shoulder blades (much like his mother did when he was with her) and shaking it a little) will not tell teach your dog that you are the boss. Rather, this sort of aggressive behaviour will only make your dog think that you are a dangerous creature who should be avoided or taken down.

In a position paper released by APDT in 2009, the group stressed that physical or psychological intimidation hinders effective training, and damages the relationship between humans and dogs. APDT advocates training dogs with the emphasis on "... rewarding desired behaviours and discouraging undesirable behaviours, using clear and consistent instructions and avoiding psychological and physical intimidation." The association pushed for the adoption of modern, scientifically-based dog training that emphasises teamwork and a harmonious relationship between dogs and humans.

The American Veterinary Society of Animal Behaviour (AVSAB) also issued a position paper junking the idea of using aggression to teach an animal to submit to his owner's will. AVSAB said owners should strive for leadership by example and not dominance when it comes to handling their dogs.

APDT noted that its trainers and animal behaviourists have already come up with concepts that emphasise the need for owners to build a happy and healthy relationship with their dogs, instead of using aggression to train them. Trainers have begun to recognise the importance of helping owners find non-confrontational and more humane ways to deal with their canine companions.

Some trainers employ programmes such as Nothing in Life is Free (NILIF), which is based on the principle that the dog must 'do' something or complete a task before he gets what he wants. This is how it works: using positive reinforcement techniques, teach your dog a few commands or tricks. Two very useful commands are 'sit,' and 'down;' tricks may include 'roll over' and 'shake.' Once your dog is familiar with the tricks and the commands he has learnt, you can begin practising the NILIF method. In exchange for food or some attention, ask your dog to perform one of the commands you taught him; for example, ask that he lay down and remain in that position until his food bowl has been placed on the ground. The American organisation, Sacramento Society for the Prevention of Cruelty to Animals (SSPCA), advises that once a command has been given, it must be carried out before you give your dog his food (or attention). If he does not comply, simply walk away at first, returning after a few minutes to try again. Eventually, your dog should realise what is expected and do as you ask; on no account should you lose your temper, or shout at him for not obeying. It's usually the case that this only happens because the dog has not been properly taught what he should do. NILIF works, SSPCA said, because it effectively and gently communicates to your dog that, as you have access to and control of the valued resources such as food and playthings, you are the leader. Owners are advised to encourage children to practice NILIF with their dog to establish that they are not playmates, simply because kids are small and get down on the dog's level to play.

Clicker training is another method that is considered humane. It involves the use of a small, hand-held device that emits a distinctive click when pressed. When a dog does something right, the clicker is clicked and a treat given, so that the dog comes to associate the noise of the clicker with a good thing. (In time, you will be able to dispense with giving a treat on every occasion as the click on its own will be sufficient to let your dog know that he has behaved as required.) The Humane Society Silicon Valley (HSSV) noted that, for more than half a century, the method has been successfully used in training over 140 different species of animals worldwide. HSSV pointed out that clicker training zeroes in on teaching acceptable behaviours using strong positive reinforcement, and approves of its use for training small animals. Interestingly, most of the animals that have appeared in films and television commercials in the US have been trained using the clicker method.

Some animal welfare groups such as HSSV are opposed to the use of choke collars. Citing medical studies, they contend that choke collars can cause trachea damage: when used with force, a choke collar can cause a dog's neck or back vertebrae to go out of alignment. Aside from causing pain, HSSV says that choke collars can increase aggression in a dog.

If you have more than one dog, it's always advisable to separate the dominant dog from subordinates during feeding time. Also, when giving treats or special attention, you should not commit the mistake of calling the subordinate dog first. More than likely, the subordinate dog will not comply in any case, since obeying you would mean disregarding pack hierarchy.

The Humane Society of Missouri noted that puppy training should begin as soon as they are brought home. Young puppies of seven to eight weeks of age can begin to learn simple obedience

CLEVER DOG!

commands such as 'sit' and 'down,' employing methods that use positive reinforcement and gentle teaching.

The earlier you begin training your puppy, the less likely it is that he will develop behavioural problems. It's known that once a puppy reaches the age of 6 months, he will be much more difficult to train and socialise, since adult behavioural patterns and dominance behaviour may already be evident.

Pack mentality will come into play when training a puppy. As a new member of the 'human pack,' your young friend will have to learn what his rank is within the pack. For instance, your child may approach the puppy whilst he is playing with a toy. Since it is your dog's prized possession, he may warn the child not to come closer by growling.

According to veterinarians and animal behaviourists, the NILIF method is a good technique for training puppies. Your young friend should learn early on that 'nothing in life is free,' and that to get what he wants, he must earn it.

The use of a reward such as small pieces of food, or a favoured toy is advisable, according to animal advocate groups. You will get the desired response from your puppy if your treat or toy proves appealing to him. Show the puppy the treat or reward, then give your command; by moving the reward you can show the puppy what is required. For example, with your dog standing in front of you, hold the reward just above his nose. and begin slowly moving it backwards over his head. As he follows the treat with his eyes, his bottom should sink to the floor; as it does so, say 'sit!' and give the reward, praising him at the same time. Likewise, for a 'down!' command, move the hand with the treat down to the floor just in front of his nose, and hold it there until he is in the correct position. Again, say the command and give the treat, as well as much praise. The idea is to pair a command phrase or word with each action and give the reward for the appropriate response.

Training puppies and adult dogs entails repetition, time, perseverance, and a great deal of patience. Consider training classes if you are unsure about carrying out the training yourself, but be prepared to put into practice what is taught during lessons. Some veterinarians recommend this as trainers can help you resolve puppy training problems. Also, your dog will have an opportunity to become socialised to both people and other dogs in a controlled environment. In training classes, you will be able to immediately detect potential behavioural problems and correct them before they become unmanageable.

Veterinarians and animal behaviourists are firmly of the opinion that puppies educated using humane training methods, which emphasise the need to submit to you so they can get what they want, are more likely to grow into well-adjusted, happy dogs.

In dealing with your dogs, the most important thing you have to keep in mind is to provide him leadership, to act as his leader. Once your dog recognizes your leadership, he will relax and fall easily into the role of your family's pet.

Lessons we can learn from dogs

When it comes to leadership, dogs are an extremely good example of how to do it right. Once a dog establishes itself as leader, it will hold that position for life. In our world, leaders are ousted, they self-destruct, they back away from the challenge, and they let down their followers. Too often human leaders say one thing and do another – that's the crucial difference between human leaders and canine leaders.

There is nothing a dog would expect members of his pack to do that he hasn't done himself. He is the leader because the pack trusts him. When you bring home the shopping, your dog sees you as the provider of sustenance. He thinks you've been out hunting and he's always impressed with the bounty. If we can show those we ask to trust us that their trust is always going to be repaid, we'll never have to prove ourselves.

Don't promise what you don't know you can deliver. Don't try to oversell yourself, people can see through false claims. Dogs don't try to impress; they get on with what they need to do, whether it's finding food, protecting the pack, or getting rid of rats. They just do it.

FOUR

HOW DOGS BENEFIT HUMAN HEALTH

Man's best friend, Man's best doctor

"To sit with a dog on a hillside on a glorious afternoon is to be back in Eden, where doing nothing was not boring – it was peace." – Milan Kundera

It's ironic. So keen was he to prove that dogs were a common cause of allergies in children, he insisted that any child growing up with a dog was likely to spend more time off school, and sniff and sneeze their way to adulthood. Egg truly landed on face when, in fact, it was proven that the exact opposite is actually the case.

I refer to a robust exchange I once had with an adamant young man who was preparing a radio phone-in on the dangers of dogs and children. He'd decided/assumed that not only did dogs represent a safety hazard to youngsters, they were also (probably) the cause of all manner of allergies and skin complaints, and all but totally responsible for just about every possible ailment a child could suffer in their formative years.

Whoops.

He couldn't have been more spectacularly wrong.

If you are the parent of a young child with asthma, you will be well aware of the differing views about whether or not it's a good idea to have a pet in the house. As it turns out, at the time of the phone-in, I was already aware of the six-year study of 9000 children that had been carried out in Germany, which actually found that children have less chance of suffering from allergies if they share their home with a canine member of the family.

Professor Guy Marks, head of the epidemiology group at the Woolcock Institute of Medical Research in Sydney, conducted the research, and couldn't have been clearer in his interpretation of the findings when he told ABC News Australia:

"This research is another piece of evidence which fits with previous research to suggest that there's something about having pets which seems to be associated with protection against development of allergic disease. There have now been a number of these studies. Most of them have related to cats, but some, like this one, relate to dogs.

"I suspect there's nothing specific about pets, about cats and dogs, it's probably a general phenomenon about having household pets.

"What it is about having the pets that's protective still remains unclear."

When asked whether the study findings gave credence to the theory that children in non-dog households were simply unable to build up immunity as a result of an all but complete lack of 'access to bacteria,' he responded:

CLEVER DOG!

"That's been one of the interpretations of this type of evidence – that not having pets around the house is associated with a cleaner environment, and that that increases your risk of having allergy."

One of the key findings of the study was that, in order to assist in helping youngsters build allergy immunity, a dog had to actually live in the home with the child, not just come to visit, or be in regular contact with the child. So, in telling my radio researcher that he had got his theory upside down, I needed only to quote Professor Marks:

"What I think it does [the research] is free parents to make up their own minds on other grounds, as to whether or not they want to have pets."

Professor Marks' study isn't the only one of its kind to reach similar conclusions.

Research from the University of Warwick in the UK showed that, when tested, the saliva of 138 children indicated that enhanced immune system strength was higher in children who had animals in the home. Furthermore, the animal-owning children were taking on average nine fewer sick days a year.

Children growing up with dogs are less likely to suffer allergies. Another of the many examples of how our best friend contributes to our lives in a positive way.

The notion of a dog being good for our health has moved beyond the anecdotal evidence of – admittedly biased – dog owners, and is now officially recognised, thanks to scientific studies which show 'dog power' can help us live longer, recover more quickly from illness, and be more productive at work, so the suggestion that dogs are the cause of disease and uncleanliness, or are detrimental to human health, has always been a hot button for me.

I remember sitting in a lecture room during a conference on canine cancer. As scientist spoke to scientist I have to concede that much of what was said went over my head. Then, an American gentleman stood up to give a presentation on the use of the controversial drug thalidomide in the treatment of oral cancer in dogs. Even though I didn't take notes (I tried, but gave up after the use of the umpteenth word I didn't know the meaning of), I distinctly remember the thrust of this particular presentation.

We were told that dogs will respond to treatment or fail to respond to treatment in the same way that humans do. So if a drug has a positive effect on managing oral cancer when it is used on a dog, it will have a similar effect when used in humans. Likewise, if a drug proves ineffective at fighting oral cancer when used on a dog, it will likely show the same lack of success when used in humans. This seems quite logical but, for some reason, I'd never really given the topic the thought it deserves. Essentially, the medical treatment of dogs acts as something of a model for the treatment of the same ailment or disease in humans.

During the presentation we were told about an elderly dog that had been diagnosed with oral cancer. His owner was distraught. The prognosis, bleak. Asked if she would be willing to consider an experimental treatment using thalidomide, she willingly agreed – after all, the stark choice between that or a fairly certain, rapid decline made the decision easier.

The dog began the treatment. After a couple of weeks, there was no improvement, and the situation remained the same a further two weeks later. Two months in, no real change. At this point in the presentation our lecturer informed us that, had this been a human undergoing this experimental treatment, it would have been halted there and then. Ethically, the medical profession would not be allowed to continue for too long with a particular form of treatment when there was no improvement. Instead, care would change to the management of the disease, which really means administering painkiller until the inevitable occurred.

In the case of the elderly dog, after nine weeks of treatment with thalidomide, something incredible happened; the cancer began to go into rapid remission, and, by twelve weeks, had all but gone. After four months of treatment, an animal whose owner was originally facing the prospect of euthanasia for her beloved companion, was totally well and free from what had been diagnised as terminal cancer.

Why this story is so incredible, we were told, is that it established a principle that would never have been arrived at were it not for the treatment administered to that dog; that thalidomide works, but only after it has been used for at least two months. Furthermore, expecting small, early changes was pointless; the changes do happen, but usually only after two months and then they are rapid.

Could dogs hold the key to predicting cancer?

Sadly, over a quarter of domestic dogs will die of cancer. A similar number of domestic cats are also prone to specific malignancies. Current treatment of these diseases is improving, but the success rate for cancer therapy remains poor. With any malignancy, early detection is essential for a good outcome. However, most pet cancers are detected late, which results in a poor prognosis for the animal. As more of us seek to prolong the lives of our dogs, is it possible we will actually see the future of cancer treatment emerge from the veterinary world?

Dr Kevin Slater of the specialist canine cancer screening firm PetScreen explains how scientists are working on improving the success rate of all cancer treatments.

"PetScreen is a company that aims to develop innovative tests to improve cancer diagnosis and prognosis in companion animals.

"The company was founded by people who have lost their pets to cancer, and by scientists who have had years of experience working on new methods used in the study and treatment of cancer. Our goal is to use the latest scientific developments to make animal cancer therapy kinder and more effective."

As PetScreen is about to begin clinical trials relating to cancer care for animals, Kevin is only too aware of what an emotive issue he is dealing with. Although the pet industry as a whole is waiting with bated breath to see the findings, he is quick to caution that predictive health screening is still in its infancy.

However, there's a firm set of objectives in place to begin the challenge of spotting and treating illnesses before they actually take hold. The first objective is one of those ideas which, once you take time to consider it, evokes the phrase 'ingenious, yet simple.'

Kevin explains:

"In the UK and Europe, animal cancers are usually treated by surgery. However, chemotherapy, when properly administered, can provide significant improvement in cancer care. We aim to develop advanced techniques that facilitate tailored (directed) therapies for individual animals. The technology, called the Tumour Chemosensitivity Assay (TCA), will be offered to veterinary practices as a service."

This process involves the vet taking a biopsy from the animal suffering with cancer and sending it to the PetScreen lab. The process then involves the tumour being allowed to grow in the lab under exposure to many different treatments, permitting PetScreen scientists to observe which treatment is working best for any particular tumour.

Although at this stage this is only an objective not yet attained, it's easy to see the logic of it. Developing the process as a service available to vets should improve the recovery/survival rate by making treatment for individual animals more effective.

The second objective relates more to the idea of early detection expressed earlier. Usually, animals do not communicate illness to their owners until the disease is well developed, meaning that most cancers are detected late, when treatment is more difficult.

"We will develop early diagnostic markers for cancer based on emerging 'proteomic' technology (specific proteins are present in high concentrations in the blood, and the pattern of these proteins can be analysed and used as a 'fingerprint' to identify a tumour). These tests will be offered on an annual routine screening basis (at the same time as the animal's annual booster vaccinations) to detect cancer at its earliest possible stage. Treatment of cancers detected early in their development is dramatically more successful than if the tumour is found at a late stage. Currently, the majority of cancers in companion animals are detected late, which can result in devastating consequences for the pets and their owners."

What we can learn from the PetScreen approach is that a revolutionary method of early

CLEVER DOG!

detection is viable. When a cancer has been found at an early stage, treatment can be much more effective. Combine this with the tumour chemosensitivity assay, and there is a distinct possibility of cancer care being kinder and less traumatic for animal patients. Not only could detection be more effective, the need for repeated and frequent treatments such as chemotherapy could be a thing of the past.

The development of veterinary medicine has been making outstanding progress over recent years, so much so that the life expectancy of our pets gets longer and longer. With this comes the realisation that, as there are more elderly companion animals, there will be a correspondingly higher incidence of illness. Every owner should take a more detailed interest in the health of their animal, regularly checking them for anything untoward. We have got into the habit of performing checks on ourselves, for example for breast or testicular cancer. Our pets don't have that option, so it's vital we do it for them.

This doesn't just apply to these areas of our pets' anatomies, of course; owners should carry out a weekly inspection of their dog's entire body, checking for lumps and bumps, as well as monitoring stools, and noting frequency of urination and defecation. It's also a good idea to inspect the gums and eyes, paying particular attention to any discolouration found. In this way, an owner can assist with the process of early detection.

Regarding prevention, certain simple and obvious steps can be taken to minimise risks where health is concerned. Lifestyle choices can affect the probability of certain illnesses. As with humans, if a poor diet is combined with little or no exercise, heart disease in our animals becomes more likely by the day.

"In America, it is a lot more common to treat pets with chemotherapy, whereas in Europe this is relatively rare. We are ten years behind them [the USA] in terms of how we treat cancer, but when PetScreen realises its goals that trend will be fully reversed and we will be ten years more advanced," enthuses Dr Slater.

Using the dog as a potential model for understanding the best predictive approach to cancer in humans is one thing, but they've got another trick up their sleeve. Meet the dogs who can actually detect the presence of cancer in humans.

Cancer and Bio-detection dogs

Andy Cook MA (CANTAB: Cambridge Neuropsychological Test Automated Battery) is a dog trainer, who worked on the pioneering canine cancer project run by the charity Hearing Dogs for Deaf People. Andy is famous for his work with a rather amazing Cocker Spaniel called Tangle. But before we get on to Tangle, Andy explains the background to how it was discovered that dogs could detect cancer in humans.

"It has been known for many years that dogs are capable of detecting cancer in their owners. Gill Lacey, a colleague of mine at Hearing Dogs, has first-hand experience of this. Back in the 70s, her pet Dalmatian, Trudii, persistently fussed at an apparently normal mole on Gill's leg. Gill ignored this behaviour for weeks, as the mole didn't appear inflamed or enlarged in any way and, after all, what person in their right mind seeks medical help on the advice of their dog?"

When Trudii's fascination with this mole reached the point where she attempted to nip it off with her teeth, Gill bit the bullet and consulted her GP. He, in turn, could see no reason to be suspicious of the mole, but agreed to remove it and send it for analysis for everyone's peace of mind. Tests proved this to be a clear case of malignant melanoma, an extremely aggressive type of skin cancer. Gill knows she's lucky to be alive today, thanks solely to Trudii and her self-taught diagnostic skills.

A team in Buckinghamshire – which became known as Cancer and Bio-detection Dogs – decided to take this further by examining the phenomenon scientifically, setting out to prove the principle that dogs could detect cancer without ever meeting the patients involved. Not only that, but that they could key into scents associated with specific types of cancer, rather than the disease in general, and prove that they were not merely indicating the presence of blood, pus, or inflammation – after all, any self-respecting dog would be fascinated by those scents, wouldn't he?

The Buckinghamshire team brought together a partnership between hospital researchers, clinicians and dog trainers. The latter, all working at the charity Hearing Dogs for Deaf People, trained six of their own dogs in their spare time, over a period of seven months.

"I was privileged to play my part by training one of the dogs, a little chocolate Cocker Spaniel bitch called Biddy," explains Andy: "My wife, Claire Guest, trained Tangle, another chocolate Cocker Spaniel from field trial lines, and two other work colleagues, Sandra Stevenson and Jan Smith, trained two dogs each, including a Labrador, a mongrel, a Papillon, and another Cocker Spaniel. Amersham Hospital supplied us with urine samples from bladder cancer patients and from other patients with a variety of non-cancerous urological diseases, which the dogs would need to learn to ignore.

"During the advanced phase of training the dogs consistently and persistently indicated a supposedly cancer-free specimen. We became very demoralised because we thought they were unable to tell the difference between someone who was very ill with an illness other than cancer and someone who actually had cancer. To be on the safe side, the hospital decided to call the patient in for further tests, and it was discovered that he did indeed have cancer in one of his kidneys. Not only that, it was also bladder-type cancer – the very type the dogs had been trained to detect. The hospital was obviously impressed and the patient's life was saved."

Tangle, a small, liver-coloured Cocker Spaniel, began detecting cancer as part of a trial performed on behalf of the British Medical Journal, and is the first dog in the world to be put to work in such a capacity. After registering a 100 per cent success rate in the initial clinical trials, efforts were made by the charity Hearing Dogs For Deaf People, where he was initially trained as a hearing dog, to ensure that his special talent did not go to waste.

Since completing the trials, Tangle's accuracy at detecting and signalling the presence of cancer has become more acute. This amazing canine needs only a half millimetre sample of urine to detect the presence of cancer.

Andy explains: "Tangle can tell the difference between a urine sample that's been given by a patient with bladder cancer from other urine samples that have been given by patients, many of whom have urological conditions like kidney stones, but do not have cancer. He does this by smell and indicates the sample which he believes to be positive by lying down in front of it."

Dogs can detect cancer: what are the implications?

A team of dogs successfully proved that they could detect cancer even when put through a fully blinded clinical trial, and their results were so statistically significant that the study was published in the *British Medical Journal*. We can now say with certainty that dogs have the ability to sniff out cancer in humans, an astonishing breakthrough in our ability to understand and recognise that dogs have talents which we haven't yet fully harnessed. Where could this lead?

Other types of cancer, certainly, and possibly other diseases, too, could be detectable by dogs, if we train them in the correct way. We already know that many pet dogs indicate changes in blood sugar levels to their diabetic owners, and that dogs can predict imminent epileptic seizures. Could we be looking at a future where it is entirely possible that dogs will be trained to diagnose a whole variety of medical conditions, in humans and other animals? Can we conceive of a world where people will have their own medical detection dogs? Their own private doctor on four legs?

Andy Cook is excited at the prospect: "The only limiting factor is our imagination and our access to patients, for which we rely on collaboration with open-minded clinicians such as our own team at Amersham Hospital. Whether individual dogs can be trained to detect a variety of diseases – doggy GPs – or whether they will specialise in one disease alone, remains to be seen. We intend to continue researching this area and are determined that we will continue to push back the boundaries of knowledge over the coming years."

Can dogs help humans avoid cancer?

Incredibly, dogs not only have the ability to detect the presence of cancer in humans, but,

CLEVER DOG!

according to research, their very presence can help us avoid the deadly disease. Dog owners are thought to be at least a third less likely to contract non-Hodgkin's lymphoma, which affects around 9000 Britons a year, say scientists at the University of California, San Francisco, and Stanford University. The scientists who carried out the study believe dogs can help protect against cancer by boosting the immune system of their human companions.

Non-Hodgkin's Lymphoma affects the lymphatic system, a network of vessels and glands that transports infection-fighting white blood cells round the body. Cases of NHL increase with age; the average age of newly-diagnosed sufferers is 65, and the most aggressive form of NHL can be fatal, which further illustrates how monumentally impressive it is that simply having a dog in the home can reduce the risk by even a tiny percentage, let alone a third.

Evidence suggests exposure to allergens and toxins in the environment may actually help protect the human body from certain types of cancer.

In the latest investigation, researchers at the University of California, San Francisco, and Stanford University, also in California, studied more than 4000 patients to see if owning a dog affected their chances of contracting cancer. Just under 1600 of the volunteers developed NHL, while the remaining 2500 or so were free of the disease. Of these, results showed that dog owners were almost 30 per cent less likely to develop cancer than those who had never kept animals. The longer the family had kept pets, the greater the protection against the disease, the study found.

In a report published in the journal *Cancer Epidemiology, Biomarkers and Prevention*, researchers said: "This large study provides support for the theory that pet ownership decreases the risk of Non-Hodgkin's Lymphoma. And it's possible this is related to altered immune function."

Dr Jodie Moffat, health information officer at Cancer Research UK, told Britain's *Daily Mail* newspaper: "This study presents some evidence that owning a pet may reduce the risk of Non-Hodgkin's Lymphoma. However, much more research is needed to provide a conclusive link.

"Scientists around the world are trying to find out what causes Non-Hodgkin's Lymphoma, and what affects a person's chances of developing the disease, but there is still a lot to learn."

DNA profiling

If someone came into your home, in broad daylight and stole your dog, you'd call the police. But what happens when you are asked to prove ownership of the dog? Well, of course, you'd reach for your documents, photographs, and all the other bits and pieces that 'prove' you own your dog. But, believe it or not, these prove nothing. The thief could have a handfull of similar documents, photos and items that proved he too owned a yellow Labrador that looks a bit like yours. Alarming as it is to realise, with only the usual paperwork, you cannot prove categorically that the dog in question is yours, only that you own an animal which 'matches the description' of the dog you call your own. Of course, microchipping provides pretty conclusive proof of ownership, but chips can be removed ...

Which is where DNA profiling comes in. Already used in paternity tests, forensic evidence, and medical research, this process is now available to dog owners.

George Clottey of Blueprint Healthcare explains how DNA profiling will benefit dogs and their owners. "DNA fingerprinting is now well established as a reliable means of identification for dogs. Currently a test accessed via the UK Kennel Club and largely used by breeders, it is a definitive way of proving pedigree.

"However, an individual dog's DNA not only defines identity, it can also reveal genetic characteristics of the dog itself. A number of tests are now available for genetic diseases affecting the dog population. Identifying whether a dog suffers from, or is a carrier of these diseases is helpful both for breed management and also for the care and treatment of the individual dog."

Over one hundred thousand dogs are lost each year, and fifty thousand are reported stolen, many of which are never reunited with their owners. Because a DNA profile is unique, it allows unequivocal identification, and because it is permanent, it can never be altered, modified or removed. Therefore, if a dog's DNA profile is registered on Blueprint Healthcare's database, it can always be identified by this. As this technology finds a willing customer base in dog owners, humans will reap the rewards of the investment and ongoing improvements in DNA technology via our canine friends.

It's currently impracticable for reasons of privacy, data protection and cost, to ask large human populations to supply a DNA profile for storage, but dogs – ever the willing partner – don't share the same concerns about privacy. Yet again, this could be another instance of learning more about the uses for DNA, as well as the potential benefits for human society, thanks to man's best friend.

The canine impact on health and well-being is no better exemplified than the relationship between an assistance dog and their partner.

Allen Parton suffered serious head injuries whilst serving his country in the first Gulf War. He was left with severe memory loss and found himself confined to a wheelchair. Somewhat reluctantly, Allen was given an opportunity to work with a young yellow Labrador by the name of Endal; little did he know at the time, but this was to be a life-changing meeting.

Endal had been trained by the British charity Canine Partners (or Canine Partners for Independence, as it was known at the time). Embarking on a journey that epitomises the symbiotic relationship between man and dog, Endal would help Allen with a wide variety of daily tasks (he even learned to operate a cash machine). The bond between the two was so strong that their story was eventually depicted on screen and in print via book and film deals.

"I could not recall getting married, or the birth of my children, and had lost about 50 per cent of my history (none of which has ever come back). My speech was awful and my behaviour bizarre. My memory then and now lasts about two days and I'm dependent on a wheelchair for mobility," says Allen.

"Because of my self-pity, anger and bitterness I was stuck in the darkest, soulless place a person can be, devoid of any hope. Without the usual emotions of love, hate, happiness and sadness, I was beyond human help, even out of the reach of my loving wife and two young children. I defended myself from people by being horrible and rude; that way I would be left alone.

"One dog cut right through that defensive armour and saw the real Allen Parton. Like a shining star, Endal came bounding into that dark place and touched my very heart. He just said to me with those doggie eyes 'Hold onto my tail and I'll pull you out of here, at your pace and with no conditions.' His unconditional love has healed so many of the hurts, his mischief-making brought laughter into my saddest days, and his zest for life has rubbed off on me.

"If I was to think of the most important thing Endal has opened my eyes to over his thirteen years amongst us, it has to be his ability to expand so many people's minds to how important and valuable dogs are in our lives," says Allen. "I still can't pick a point in my life when I can say I did something that deserved Endal's unconditional love and devotion. I have been truly blessed, but now I realise that Endal was never just my dog, he was everyone's.

"His greatest legacy has to be bringing me back to my family and also mentoring faithful little EJ [Endal Junior, Endal's successor]. I had been given the opportunity, on the morning before he left us, to tell Endal just how much I loved him, and to thank him for all he has done for me through his life; that was possibly the most important and significant moment of our entire relationship."

Allen and Endal's relationship is one of thousands enjoyed by people fortunate enough to be partnered with an assistance dog. Most people are familiar with the work of guide dogs, who assist blind and visually-impaired people, but there are now many more assistance dogs lending a helping paw to their partners. Here's just a selection of the UK-based assistance dog organisations:

Guide dogs
The Guide Dogs for the Blind Association provides dogs, mobility and rehabilitation services to blind and partially-sighted people.

Hearing dogs
Hearing Dogs for Deaf People trains dogs to alert deaf people to specific sounds, whether in the

home, workplace or public buildings. Some dogs from Battersea Dogs Home have been trained as hearing dogs.

Dogs for the Disabled
Originally part of Guide Dogs for the Blind, Dogs for the Disabled provides trained dogs that promote independence for people with disabilities, by building mutually beneficial partnerships which respect the needs of both person and dog.

Canine Partners for Independence
Canine Partners helps disabled people enjoy greater independence and a better quality of life through the assistance of specially trained dogs.

Support Dogs
Support Dogs is a charity that helps improve the quality of people's lives who may be suffering with disabilities such as epilepsy.

Pets as Therapy (PAT)
Pets as Therapy (PAT) facilitates the provision of dogs who make visits to hospitals, hospices, nursing and care homes. These dogs have a positive effect on people who may be ill or confined by physical disability. They lift spirits, don't judge, and are chosen specifically for their happy disposition and calm nature.

Medical Alert Dogs
A charity that allocates specially-trained dogs to individuals of all ages who manage complex medical conditions – such as diabetes and Addison's disease – on a daily basis. The dogs warn their owners of impending events by barking, jumping up or licking, and can also fetch any necessary medical supplies and help.

Assistance Dogs International (USA based)
Assistance Dogs International, Inc is a coalition of not-for-profit groups, training and placing assistance dogs with partners.

Whilst we all may be becoming more accustomed to seeing assistance dogs working with their human partners, more amazing relationships are being formed between man and dog every day.

Take, for instance, the Dallas-based Diabetes Friendly Foundation which operates the 'K-9 for Kids' programme, providing assistance in locating Diabetic Alert Dogs (DADs), and providing funding for the training and placement of dogs which are trained to assist people suffering from diabetes.

"Every 24 hours, 4000 people in the US are diagnosed with diabetes, according to the Center for Disease Control," said Cole Egger, Founder of the foundation. "Parents of children with diabetes constantly have to check their child's blood sugar throughout the day and night in order to prevent severe consequences caused by fluctuating blood sugar levels. With a diabetes alert dog, some of the strain and fear of the dangerous side-effects of diabetes is lifted."

If there was ever any doubt about how important these diabetic alert dogs are, just read what their owners say about them:

"We were very lucky to have Mallie," says Crystall Young, mother of 19-month-old Ean, who has Type 1 diabetes. "She truly is an amazing alert dog. She alerted me at times when I wouldn't have thought I needed to check on Ean. There were even occasions when Ean was outside playing and she would alert from inside."

Deanna Whitehead is a teenager who suffers Type 1 diabetes. This is what she has to say about JD, her diabetic alert dog.

"Without JD, my diabetes alert dog, I was afraid to do a lot of things, but he has given me the confidence to live my life now and not be afraid any more. I trust him to take care of me and it's an awesome feeling. I love JD."

Four great health reasons for owning a dog
• If you want to live a healthier life, get a dog. Research carried out by Queen's University, Belfast, showed that dog owners tend to have lower blood pressure and lower cholesterol. Writing in the *British Journal of Health Psychology*, Dr Deborah Wells, a psychologist from the University, said pet owners tended in general to be healthier than those without pets.

• Dog owners were also found to suffer fewer minor ailments as well as serious medical problems. This, coupled with the suggestion that dogs can aid recovery from serious illnesses such as heart attacks, and act as an early warning system to detect an epileptic seizure, is just another of the great health benefits of dog ownership.

• Having a dog can help children develop better social skills. Researchers at the University of Leicester found that children between the ages of birth and six in dog-owning families will have better social skills, better speech, better co-ordination, and greater confidence, and will be less likely to suffer from allergies.

• A five year study of 600 children aged 3-18 years revealed that the children of dog-owning families who are slow learners or whose parents have divorced cope better with life than those who don't have a pet.

Dogs and children: the health rewards
Young children who interact and play with dogs are up to 50 per cent less likely to be overweight or obese, suggests research from Deakin University, Melbourne. "Even incidental play with a dog helps keep the weight off," says head researcher, Jo Salmon.

Intimacy
More than 90 per cent of children list pets in their top ten most special relationships, says Dr June McNicholas, a psychologist from the University of Warwick who specialises in human-pet interaction. "In some cases, pets even came first, above all human relationships." The children, aged seven and eight, said they confided in pets and turned to them for comfort when ill.

Pets help children develop patience and problem-solving, says Maggie O'Haire from the University of Queensland's School of Psychology and Centre for Companion Animal Health.

"A dog or cat won't do what a child wants them to do all the time." This, she says, teaches compromise.

"Once a child considers an animal's point of view, they tend to change their behaviour. These lessons often translate to better communication with the people in a child's life. Learning to communicate with a non-verbal other is also a great lesson in patience."

Responsibility
"Pets introduce [children] to routine responsibility," says O'Haire. "Caring for an animal may be one of the few outlets kids have to learn nurturing and care-giving skills." This facility for care can translate to appropriate views on animal welfare as a whole. "What a child considers to be acceptable treatment of farm animals or wildlife often reflects how they'd feel if those treatments were applied to their beloved family pet," says Dr Paul McGreevy of the University of Sydney's Faculty of Veterinary Science, and author of *Handle with Care – Making Friends with Animals*.

Better understanding of health and illness
A study by McNicholas found that children with pets have a better understanding of medical procedures. "The health treatment of a pet cat may be more thoroughly explained to them than any treatment they, or a close family member, may receive," says McNicholas. Kids with pets could also be better able to handle the concept of mortality.

CLEVER DOG!

"The death of a pet can be a very sad event, but it is a way to teach kids important lessons about the life cycle," adds O'Haire. "It can create space for meaningful discussions."

Dogs: the pet with the most health benefits for humans

It has been established for some time that owning a companion animal makes people happier, with the benefits of dog ownership/co-habitation outstripping those of a cat or any other animal.

A Queen's University, Belfast, study has shown that dog owners tend to have lower blood pressure and cholesterol.

Dr Deborah Wells of Queen's reviewed dozens of earlier research papers which looked at the health benefits of pet ownership. In some cases, the research even ventured as far as to suggest that the social support offered by an animal is greater than that which another human could offer. Dr Wells confirmed that pet owners tended, in general, to be healthier than the average person, with dog owners benefiting most by suffering fewer minor ailments and serious medical problems than owners of other animals.

It has been believed for some time that dogs can assist people recover from serious illnesses such as heart attacks, as well as give early warning of an imminent epileptic seizure. This research further strengths that claim.

Dogs as stress relievers

"It's possible that dogs can directly promote our well-being by buffering us from stress, one of the major risk factors associated with ill-health." believes Dr Wells.

"Owning a dog can also lead to increases in physical activity, and facilitate the development of social contacts, which may enhance both physiological and psychological human health in a more indirect manner."

Lessons we can learn from dogs

If there was ever any doubt that dogs are good for us, this chapter surely debunks some of the anti-dog myths that abound to this day. Myths such as dogs as a 'cause' of allergies. Or dogs as inherently unhygienic or 'dirty.'

Statistics and studies show that, in fact, quite the opposite is true. Dog owners are, in the main, happier, healthier, take fewer sick days, suffer less stress, live longer, cope with difficult circumstances better and always, always have a shoulder to cry on whenever they need one.

Is it really any wonder that this magnificent animal has earned the accolade of man's best friend? I'm regularly humbled by the dogs I meet, and, if I may paraphrase from a quote I once read, "If I can be half the person my dog thinks I am, I'd be very happy." Well, I'd like to slightly alter that to "If I can be half the person my dog is, I'd be deliriously happy."

As our dogs continue to unravel medical mysteries or, more to the point, as we continue to uncover their hidden abilities as 'dogtors,' we must surely try and discover just what else is contained within the canine psyche that could hold the answers to so many questions, not just about our own health and wellbeing, but our dogs, too.

When you get right down to it, the reason why dogs have such a beneficial effect for us is that they make us happy. Many people say the thing they value most about their dog is the welcome they get whenever they arrive home. In Dale Carnegie's book *How To Win Friends and Influence People*, he talks about that welcome and asks his readers to apply it to their daily lives. Be genuinely pleased to see people. If you think about it, how many times have you left after being with a friend and thought to yourself "it was great seeing him; we should do that more often"? So why not have that thought in your head the next time you meet, and let them know from the start just how pleased you are to see them?

PROBLEM-SOLVING: THE CANINE WAY

Why we should value all types of canine intelligence

"There's an old saying: if you don't think a dog can count, put two biscuits in your pocket, give him one and see what happens ..." – John Mitchell

It's only natural that, as caring and devoted dog owners, we look for signs of intelligence in our dogs. Of course, some dogs are more intelligent than others (as is also the case with people ...), and most of us will openly concede that our mutt may be somewhat lacking in the brains department. Likewise, dog owners will, without too much encouragement, happily relate the shards of brilliance their dog demonstrates in their everyday behaviour; whether it's moving out of the path of a falling plate, or a tilt of the head at the sound of a familiar car engine.

We look for the same attributes in our dogs that we look for in our friends, family, and people we meet on the street: a sense of humour, a keen mind, loyalty and compassion. But when it comes to describing our dogs, the attribute we tend to be most impressed by and proud of is our dog's intelligence and the clever things he or she can do.

Different degrees of dog intelligence
When we contemplate dog intelligence, we often think of a dog breezing its way through complicated exercises in an obedience class. We might also think of highly trained canines, such as guide dogs, police dogs, or search and rescue dogs, performing their specific tasks in an efficient and intelligent manner. The sight of a dog attending to its master's commands and signals, whilst at the same time responding in a quick and assured manner to the task at hand, gives us the impression that we are viewing the peak of dog intelligence.

To work effectively under human direction obviously requires that a dog have at least enough adaptive intelligence (learning and problem solving ability) to figure out which behaviours are expected when he receives a particular command. Deciphering what a particular word or signal means is, from the dog's viewpoint, a problem that has to be solved. Experienced dog trainers often say that the hardest part of training dogs for competition in the higher obedience classes is simply trying to get them to understand what is expected.

When a dog demonstrates that he understands what particular commands mean by responding appropriately, he is displaying one of the most important aspects of canine intelligence; important, because if dogs did not respond to human control and command, they would not be capable of performing the utilitarian tasks that we value them for. Since these qualities of intelligence are also demonstrated in dog obedience competitions where dogs must execute learned exercises

CLEVER DOG!

according to human direction, we could easily call this dimension of astuteness 'obedience intelligence.' However, since it is also the acumen necessary to accomplish tasks in the real world under the guidance of a leader, we could equally as well call it 'working intelligence.'

A dog's problem-solving ability can be considered a higher level of intelligence than that required to learn something – a command, say – from a person. Take, for example, placing a treat under a bowl, which the dog knows to turn over to get the treat; this is problem-solving intelligence which not all breeds possess (and which can vary from dog-to-dog within breeds).

The theory: Dr Stanley Coren – The Intelligence of Dogs

"There are three types of dog intelligence: instinctive (what the dog is bred to do), adaptive (how well the dog learns from its environment to solve problems), and working and obedience (the equivalent of 'school learning')."

As you can imagine, when it comes to making objective observations about which dogs or breeds of dogs are most intelligent, it can be hard to find anyone who doesn't have a dog breed that they favour above all others. Still, someone has to do the job and one brave soul penned together a highly acclaimed and highly criticised book called *The Intelligence of Dogs*.

Stanley Coren, an American psychology professor and neuropsychological researcher based at the University of British Columbia, has become best known for a series of books regarding the intelligence, mental abilities, and history of dogs.

Intelligence has a variety of different dimensions. In human beings we might sub-divide it into verbal ability, numerical ability, logical reasoning, memory, and so on. The intelligence of dogs also has several different aspects, among which we recognize three major dimensions.

Adaptive intelligence
This pertains to learning and problem-solving ability, relating to the knowledge and skills a dog has learnt from his environment and time on earth. Adaptive intelligence can differ among individuals of the same breed.

Example
• If your dog recognises guests after just one or two visits, and barks at one but not the other, this is a form of adaptive intelligence. You can also look at how well your dog understands the laws of cause and effect strictly by observation. Did your dog have a bad encounter with one of the guests, or does the other happen to keep a few biscuits in his pocket when he visits?

Instinctive intelligence
This covers behaviours and skills programmed into the dog's genetic code. Some breeds were bred to do a certain job, which they have carried out for hundreds of years. For example, certain dogs were bred to herd; their ability to round up animals (usually sheep), keep them close together, and drive them in a particular direction is inborn, and requires human intervention only to keep it under control and give it direction.

Example
• Sheepdogs are consummate herding dogs, possessing an innate ability that has little to do with training, though some refinement is necessary. The Border Collie is an energetic and highly responsive breed that has the special combination of instinct and training for herding and protection. Its quick mind and herding instincts make it easy to train, and it is indubitably the prime example of inherent learning skills rooted in genetic code.

Working/obedience intelligence
This form of intelligence has to do with how well an animal can follow commands; a skill and ability that can vary quite considerably, depending on breed.

Example
• A handful of different breeds are commonly used for guiding the blind, police work, or search and rescue. All Golden Retrievers have the same instinctive intelligence and calm, caring characteristics, yet, while most are quite bright, occasionally, there will be one that appears totally clueless, and would therefore be unsuitable to serve as a guide dog.

The conclusions
Coren used 'understanding of new commands' and 'obey first command' as his benchmarks for determining canine intelligence. He surveyed over 200 dog trainers in the USA and Canada, and compiled a list of dog intelligence by breed from the results. While this method of determining dog intelligence is acceptable for the fields of training and working with dogs, it does not apply to adaptive intelligence, which can be measured by ingenuity and problem-solving of day-to-day situations and difficulties.

The drawback of this rating scale, by the author's own admission, is that it is heavily biased toward obedience skills (eg working or guard dogs) rather than measuring creativity or problem-solving that is demonstrated outside of the obedience circle by other breeds of dog (eg hunting or terrier dogs).

Results of Coren's survey
The brightest dogs
Understanding of new commands: fewer than 5 repetitions
Obey first command: 95 per cent of the time or better
 1 Border Collie
 2 Poodle
 3 German Shepherd
 4 Golden Retriever
 5 Doberman Pinscher
 6 Shetland Sheepdog
 7 Labrador Retriever
 8 Papillon
 9 Rottweiler
10 Australian Cattle Dog

Excellent working dogs
Understanding of new commands: 5 to 15 repetitions
Obey first command: 85 per cent of the time or better
11 Pembroke Welsh Corgi
12 Miniature Schnauzer
13 English Springer Spaniel
14 Belgian Shepherd Tervuren
15 Schipperke/Belgian Sheepdog
16 Collie/Keeshond
17 German Shorthaired Pointer
18 Flat-Coated Retriever/English Cocker Spaniel/Standard Schnauzer
19 Brittany
20 Cocker Spaniel
21 Weimaraner
22 Belgian Malinois/Bernese Mountain Dog
23 Pomeranian
24 Irish Water Spaniel
25 Vizsla

CLEVER DOG!

26 Cardigan Welsh Corgi

Above average working dogs
Understanding of new commands: 15 to 25 repetitions
Obey first command: 70 per cent of the time or better
27 Chesapeake Bay Retriever/Puli/Yorkshire Terrier
28 Giant Schnauzer
29 Airedale Terrier/Bouvier des Flandres
30 Border Terrier/Briard
31 Welsh Springer Spaniel
32 Manchester Terrier
33 Samoyed
34 Field Spaniel/Newfoundland/Australian Terrier/American Staffordshire Terrier/Gordon Setter/
    Bearded Collie
35 Cairn Terrier/Kerry Blue Terrier/Irish Setter
36 Norwegian Elkhound
37 Affenpinscher/Silky Terrier/Miniature Pinscher/English Setter/Pharaoh Hound/Clumber Spaniel
38 Norwich Terrier
39 Dalmatian

Average working/obedience intelligence
Understanding of new commands: 25 to 40 repetitions
Obey first command: 50 per cent of the time or better
40 Soft-coated Wheaten Terrier/Bedlington Terrier/Fox Terrier (Smooth)
41 Curly-coated Retriever/Irish Wolfhound
42 Kuvasz/Australian Shepherd
43 Saluki/Finnish Spitz/Pointer
44 Cavalier King Charles Spaniel/German Wirehaired Pointer/Black and Tan Coonhound/American
    Water Spaniel
45 Siberian Husky/Bichon Frise/English Toy Spaniel
46 Tibetan Spaniel/English Foxhound/Otterhound/American Foxhound/Greyhound
    Wirehaired Pointing Griffon
47 West Highland White Terrier/Scottish Deerhound
48 Boxer/Great Dane
49 Dachshund/Staffordshire Bull Terrier
50 Alaskan Malamute
51 Whippet/Chinese Shar Pei/Wire Fox Terrier
52 Rhodesian Ridgeback
53 Ibizan Hound/Welsh Terrier/Irish Terrier
54 Boston Terrier/Akita

Fair working/obedience intelligence
Understanding of new commands: 40 to 80 repetitions
Obey first command: 30 per cent of the time or better
55 Skye Terrier
56 Norfolk Terrier/Sealyham Terrier
57 Pug
58 French Bulldog
59 Brussels Griffon/Maltese
60 Italian Greyhound
61 Chinese Crested

62 Dandie Dinmont Terrier/Petit Basset Griffon Vendeen/Tibetan Terrier/Japanese Chin/Lakeland
    Terrier
63 Old English Sheepdog
64 Great Pyrenees
65 Scottish Terrier/Saint Bernard
66 Bull Terrier
67 Chihuahua
68 Lhasa Apso
69 Bullmastiff

Lowest degree of working/obedience intelligence
Understanding of new commands: 80 to 100 repetitions or more
Obey first command: 25 per cent of the time or worse
70 Shih Tzu
71 Basset Hound
72 Mastiff
73 Beagle
74 Pekingese
75 Bloodhound
76 Borzoi
77 Chow Chow
78 Bulldog/Basenji/Afghan Hound

The drawbacks

It might seem logical to assume that dogs with the best working and obedience abilities are the most
intelligent breeds; however, it does not necessarily follow that they are the breeds which are best at
problem-solving. Many dogs with very high adaptive intelligence seem to be relatively unresponsive
to attempts at teaching them obedience exercises; the simplest obedience commands may leave
them completely at a loss. On the other hand, some dogs with only moderate levels of adaptive
intelligence can, with the right form of training, execute obedience work quite well, and even
perform apparently quite complex tricks and exercises.

There are similar problems with other ways of trying to measure canine intelligence. Individual
breeds possess very different inherited behaviours, which can affect their reaction when trying to
measure intelligence. Dogs are pack animals, of course, and, as such, they generally have the ability
to work together to hunt prey, displaying group problem-solving skills. However, some dogs have
exhibited these skills when acting alone.

Take, for example, the Lakeland Terrier, number 62 on Coren's list. This breed originated in the
Lake District of England, and, in common with many terriers, his main objective was to hunt vermin.
For generations, the dog has been used in the Lake District for the purpose of exterminating the fell
foxes which raided the farmer's sheep fold during the lambing season. Whereas most terrier breeds
have only to bolt their quarry, or to mark it by baying, the Lakeland must be able to kill the foxes in
their lair.

Terriers are renowned for being friendly, bold, and confident. They are intelligent and
independent-minded, especially when going after prey. They are rated highly for problem-solving,
as they often have to do their job alone. A terrier going to ground after a fox, badger or rat has to
decide which tunnel to take, when to pursue, and when to back up, all without benefit of human
instruction or support.

Most terriers exhibit 'selective deafness' on occasion when their attention is elsewhere,
and are not as easy to train as some breeds, such as Collies and Labradors. These dogs are not
unintelligent, they just demonstrate their brightness in ways that are not easily measurable.

CLEVER DOG!

Invaluable assets

Most dogs rarely have to deal with extremely complex tasks, and are unlikely to learn relatively complicated activities (such as opening doors) unaided. Some dogs (such as guide dogs) are specially trained to recognise and avoid dangerous situations, and must learn a considerable number of commands, understand how to act in a huge variety of situations, and recognise threats to their human companion, some of which they might never before have encountered.

It's in their nature – problem-solving

"We all want insight into how our furry companions think, and we want to understand the silly, quirky and apparently irrational behaviours [that] Lassie or Rover demonstrate." – Stanley Coren

Many people who object to Coren's dog breed intelligence table believe that canine intelligence should not be based upon human standards, and that each breed or even individual canine may possess their own unique talent when it comes to problem-solving.

Some dogs learn faster than others; even puppies in the same litter will learn at different rates. Adult canines train their young by correcting them when they behave in an unacceptable way (biting too hard, eating out of turn, and so on), and reward them for acceptable behaviour (playing with them, feeding them, cleaning them, etc). Dogs are capable of learning, remembering, solving problems, and benefiting from their experiences of life.

You could have a very intelligent dog who learns things very quickly, and who is bored by classes that emphasise too much repetition. In the case of such a dog, inspiration could come from a more creative approach to training such as stimulating latent canine instincts. For example, hiding food in a food puzzle; in this way, the reward comprises two elements – food and mental stimulation.

At home your dog may be very smart, but that doesn't mean he always does what you want him to. In fact, intelligent dogs can find themselves in a lot of trouble at times, especially if they are good at problem-solving.

When problem-solving goes wrong

Some breeds of dog are known for their stubborn nature, and may not succeed in learning (or choose to ignore) human commands. Here, I'm thinking of the Siberian Husky, Shar-Pei, and Lhasa Apso, whose aloof and independent natures may mean they are not particularly interested in pleasing their owners, though are still good at problem-solving tasks which often have a direct benefit for them (just how to get the top off the biscuit tin!).

Siberian Huskies are often fascinated with the myriad possibilities of escaping from houses and gardens, and catching small animals, often figuring out their own numerous inventive ways of doing both.

Coren was quick to note that intelligence isn't always an indication of personality, and that the smarter the dog, the greater the chance of mischief, sometimes. He says: "While a smart dog will learn everything that you want it to know, it will also learn everything that it can get away with."

We have all heard numerous tales about dogs unbolting a door, opening the fridge, or escaping out of a small gap in the fence. These activities – undesirable, as far as we are concerned, and which often land the dog in hot water – are a result of the animal using adaptive intelligence to achieve what he wants; be it a tasty treat or a stroll around next door's garden ...

To what extent are dogs guided by instinct?

Is it possible to train dogs to disregard their natural instincts and obey without question?

The pertinent dictionary meaning of instinct is as follows: "The natural impulse, apparently independent of reason or experience, by which animals are guided." This sounds very sensible. Take a puppy: every time he is asked to do something he doesn't want to do, or if he fears the approach of a bigger or fiercer dog, he quickly lies on the floor with his legs in the air and his tummy exposed to the enemy in a gesture of complete submission.

This response has been seen in domesticated dogs, generation after generation, yet it is the echo of an instinct of the wild, when, if a young puppy assumed this position, it would have deterred attack. It's a real nuisance in training when a dog does it, however, and if you attempt to put his lead on, he simply lays on his back and waves his legs in the air. It's important, then, that we train our dogs to understand that assuming this position will not save them from having to do as asked!

Dogs are greatly guided by instinct, but a whole lot more by smell, and smell can vary considerably with individual people, according to their state of mind. For example, why is it that dogs take instantly to some people and won't go near others, especially when those they don't wish to know want to be friends with them? I think each person has a particular smell – maybe an unfriendly one, in some cases – which dogs can always detect. Fear, I believe, sends out an unpleasant smell; dogs sense nervous or anxious people yards away.

Why do dogs sniff each other's bottoms? Simply to establish which pack they belong to by the scent from the anal glands. Why do dogs roll in obnoxious substances? Because in the days of wild dogs, by leaving their own scent on something, it was a way of letting their enemies know they were about, which is also why male dogs (and some females) lift their legs and urinate where another has already scent marked.

A dog must do as he is told without question, providing that what is asked of him is fair and reasonable. For example, I do not think jumping through fire is fair or reasonable, and I hate to see animals made to do it as a trick. Animals have an instinctive fear of fire from the days of forest fires, and to train them to do this trick must involve a certain amount of cruelty.

Wherever possible, when training a dog use his natural instincts if they can be guided into the right channels. Take tracking, for example. A dog's natural instinct is to find food by using his nose, and most trainers agree that food at the end of the trail is a great incentive when teaching a dog to track.

Do dogs learn by trial and error?

The belief that dogs learn by trial and error presumes they have the mental ability to link elements through experience, trying out a way of approaching a situation and then recording (remembering) whether or not it was successful.

It's then assumed that, in a similar situation, they can recall their first experience and opt for a different approach if looking for a higher dividend, or if the previous course of action proved unsuccessful. This theory presupposes that dogs, like humans, have the ability to deduce and make choices, and can project into the future to predict a possible outcome based on a previous experience.

Dogs perceive everything through the prey instinct, and can only respond to stimuli that are of relevance to this instinct. Therefore, problem-solving for a dog has to do with ascertaining whether something is pertinent to this means of perceiving and experiencing. This basic information is what dogs are after when they scent.

There is so much in man's world that is confusing to a dog. If we witness a dog trying out several different approaches before finding one that works, or giving up altogether, it isn't that he's practising. In his first impressions of a situation, he perceives several variables that aren't connected, which dilutes his ability to solve the problem. If the desire to find a solution grows, the variables eventually merge into one coherent entity, enabling the dog's instincts to take over.

By contrast, a dog that fails is exhibiting a lack of stimulation at that moment, because the elements of what he sees before him are disparate. Instead of having one problem to solve, he has many problems to deal with; the variables never come together. He tries, then stops, then starts over again without making any real progress because he's faced with a new problem at each attempt, and each time his emotional reserves are depleted further.

The dog learns through association whether or not his actions are likely to yield results, and his experiences are based on how the prey instinct has been stimulated. In cases where he shows little or no interest, it's often because the situation is alien to him, and his instincts have not been aroused. A simple example of this is that a dog will show no interest in weather changes, as when it rains, say. This is because, in the wild, the rain has little or no bearing on his existence, so nature has not programmed him to respond to it.

Do dogs experience jealousy?

A recent experiment conducted at the University of Vienna concluded that they do: I disagree.

The experiment consisted of taking pairs of dogs and getting them to present a paw for a reward. On offering this 'handshake,' the dogs received a piece of food. One of the dogs was then asked to shake hands, but received no food when he did so. The other dog continued to get the food when he did as asked. The dog that was not being rewarded quickly stopped complying, and showed signs of annoyance or stress when his partner was rewarded.

To ensure that the outcome was truly a consequence of the interaction between the dogs, rather than simple frustration at not being rewarded, a similar experiment was conducted whereby the dogs performed the task individually and in isolation, in which instance they continued to present the paw for much longer after the rewards were ceased.

Dr Frederike Range, from the Department of Neurobiology and Cognition Research at the University of Vienna, claims that this shows it was the presence of the rewarded partner which was the greater influence on the dogs' behaviour.

"The only difference is one gets food and the other doesn't; they are responding to being unequally rewarded." she says.

Researchers say that this kind of behaviour – where one animal becomes frustrated by what is happening with another – has only previously been observed in primates. Studies with various types of monkey and chimpanzee show that they react not only to seeing their partners receiving rewards when they do not, but also to the type of reward.

The dog study also looked at whether the type of reward made a difference. Dogs were given either bread or sausage, but seemed to react equally to both: Dr Range thinks this may be because they have received training.

"It's because they have to work for the reward; this confers a higher value." she said.

If we think about what jealousy is – and are logical about what we know of this emotion – we realise that it is incredibly complex, and based on a whole slew of social elements.

Human jealousy can be based on hugely different factors, it would seem. Is that due to nature or nuture? We don't really know. Even scientists can't agree a definition of what jealously actually is. What can cause jealousy can vary in individuals – we're too different and jealousy is an emotion that does not run through us all in an identical fashion. So why should it in dogs?

I absolutely do not doubt for a single, solitary second that canines display behaviour which is very easy for us to equate to jealousy. It could be displayed in acts of resource guarding, or by dogs who are particularly greedy, territorial, pack or rank motivated – but jealousy it is not. Maybe I'd quite like my boss' job, sit in his chair, work in his office, and take home his salary, but I am not motivated in this objective even in the slightest by jealousy: I simply want to do better for myself. Dogs are simply the same.

Anthropomorphism is sometimes rife. Most of the time it's harmless, and sometimes it's nothing more than us saying "I don't understand my dog but I'll compartmentalise a particular behaviour by benchmarking it against my own." This is, plainly, crazy. And it can lead to problems.

It will be better for dogs and better for us if we make a determined effort to understand them. But always, always, always begin that voyage of discovery bearing in mind one important caveat: dogs are no more human than we are simian. They are unique and we love them for it, and they are masters at making us think what they want us to, but they are dogs.

Their understanding of human body language is at a level we're not even close to: take this example ...

But they can empathise with us

A study by Birbeck College, University of London, revealed that dogs emulate human yawning, proving what many dog owners have known for centuries; dogs can and do show empathy toward their homo sapien counterparts.

"We gave dogs everything: visual and auditory stimulus to induce them to yawn," Dr Senju, a member of the research team, told *BBC News*.

The research was carried out as two tests in identical environments. In the first test a person – who was a stranger to the dogs – actually yawned; in the second test, the same person feigned yawning by simply opening and closing his mouth to determine whether or not the dogs were responding to an open mouth. The team found that 21 out of 29 dogs yawned when the person did so (an average of 1.9 times each dog), but none did when the yawning was feigned.

The researchers believe that their results are the first evidence that dogs have the capacity to empathise with humans; although the team could not rule out stress-induced yawning – but hope to in future studies.

Dogs have lived cheek-by-jowl with humans for millennia. "Dogs have a very special capacity to read human communication. They respond when we point and when we signal." said Dr Senju.

Games to play: is your dog a problem-solver?
How do I get out of here?
Please don't do this if your dog is easily agitated or anxious.
Take a large sheet, towel or blanket and gently toss it over your dog's head.
• If he frees himself from the covering in less than 15 seconds, give him 3 points
• If it takes 15-30 seconds, 2 points
• If it takes him longer than 30 seconds, award 1 point

The hidden gem
Place a treat in a square of kitchen foil and fold it twice to seal the treat.
• If your dog uses his paws to open the foil, give him 3 points
• If he uses his mouth and paws to open the foil, give him 2 points
• If he can't get the foil open and starts playing with it, or gives up trying, give him 1 point

Pick a cup, any cup
Place a treat or toy under one of three plastic cups arranged in a row. Make sure your dog sees which cup the treat or toy is under. Walk your dog in a circle for ten seconds and then let him return to the cups.
• If he goes straight to the cup with the treat under it, give him 3 points
• If it takes two attempts to find the treat, 2 points
• If he checks the wrong two first before finding the right one, give him 1 point

Time for walkies?
At a point in the day that you don't usually walk your dog, quietly pick up his lead and your keys when you know he's watching.
• If he gets excited and starts wagging his tail, award him 3 points
• If you have to walk to the door before he twigs what's happening, give him 2 points
• If he sits there with a blank gaze, give him just 1 point

Scoring
8 or more points: You have an extremely clever canine on your hands
4-7 points: Hmmm, pretty average
3 points or less: Well ... you gotta love him for trying!

Everyone loves their dog; it goes without saying, right? And they love us back just as much. However, have you ever stopped to consider the possibility that there are ways and

CLEVER DOG!

means which could enable us to love our lives with dogs even more than we do now? Unlike our relationships with our own species, we rarely consider how we could improve upon the two-way interaction between ourselves and our dogs. Why is that? Is it maybe that, because dogs are by nature incredibly adaptable and acquiescent, and because they do not (usually) do anything wrong or offensive, but do deliver the goods as far as the 'what dogs do best' front is concerned (companionship, protection, etc), we're content to let, er, sleeping dogs lie? And, in any case, provided we're forthcoming with food, water, a dry place to sleep, and the odd spot of ear-scratching, so, it would seem, are our four-legged friends.

Given our abysmal failure to evaluate our human/canine relationships, now seems an appropriate moment to offer an insight to what makes both species tick, as well as look at ways we can improve each other's lives.

Acclaimed animal behaviour expert Stephen G King explains the key motivations of the human/canine relationship from a dog's perspective.

"Our dogs spend a fair degree of their time on their own, or have, at most, one other friend to share company with. So, what happens to us and our pets in those periods between sleeping, walking and eating?

"Well, scientifically, it's an environmental event called 'enrichment,' and as we're getting scientific, here's the technical explanation: 'Environmental enrichment is the provision of stimuli, which promotes the expression of species-appropriate behaviour with stimulating activities.' Phew!

"Dogs, in their wild state, are social carnivores that can combine in packs, although they operate within three survival categories:

• Solitary predators
• Family pack hunters
• Large pack hunters

"Thousands of years of domestication and man-made selective breeding have produced breeds with modified social repertoires.

"Many dogs exhibit breed-specific behaviours that are hard-wired, like the Springer Spaniel who is visually orientated and seems to be distracted by anything that moves in the air, such as a leaf falling from a tree.

"The Border Collie likes to herd and chase joggers, motor bikes, and other things that move across the ground. Owners of animals such as these often wonder why their dog is not paying attention in an outdoor environment, despite having their favourite treats on hand which normally work so well at home ... For this type of dog, food cannot compete with a field of birds, or a herd of sheep on the move. Enrichment for them is quite specifically defined. They're telling you – almost screaming at you – what it is that makes them tick above all else, but maybe you just haven't recognised these signals for what they are."

Activity levels

Dogs spend a considerable amount of their time inactive, although, being opportunistic, are adapted to making the most of every opening, usually interested in novel items and circumstances. If a predictable and limiting environment makes non-active periods boring, a dog can become socially and emotionally lethargic.

The dog is famed for his adaptability, and a healthy adult can cope with a range of conditions, particularly if he has access to areas with different environmental stimuli.

All animals have emotionally complex lives, and need appropriate environments and stimulation in order to feel fulfilled. Good housing, giving your dog a place of his own in the home to which he can retreat, provides him with an opportunity to exercise a degree of choice; play with his toys, gnaw on a chew, interact with humans or choose not to, rest, sleep, all of which allow him to carry out natural behaviours.

Environmental enrichment for companion animals

Providing a positive, rewarding and stimulating environment based on trust and affection allows our companions to remain enthusiastic and interested.

The purpose of environmental enrichment is to increase the overall welfare of the dog by allowing him to burn off calories in a safe and natural way, which, in turn, allows more natural behaviour in a restricted environment.

The dog learns to cope with unexpected changes in the environment because he doesn't associate change with a negative outcome, and is therefore less likely to be stressed by handling or restraint. A decrease in emotion-based behaviour can result in a corresponding drop in physiological and psychological stress.

Things dogs like to work for

Food, water, sexual stimulation, foraging, sniffing/scenting, attention, grooming, coolness (when body temperature is high), and warmth (when body temperature is low). These are all known as primary reinforcers.

As these examples suggest, primary reinforcers often relate to biological processes. Some primary reinforcers are not immediately obvious; for instance, if you were a dog restrained in your home/kennel and couldn't move about or get out, the chance of freedom would be very attractive, and of far greater importance for you than for a dog who had access to open fields for most of the time.

Why is it important to know what your dog likes to work for?

Prevention is better than cure. Knowing what a dog likes and will work for is an important factor in having a stimulated dog; especially if the animal spends most of the time looking at four walls, and food is always provided regardless of its behaviour.

In the wild, dogs are hunter-scavengers; it's natural for them to spend much of their energy looking for food every day. The needs of companion dogs are similar, and certain elements of our lifestyles which our dogs must fit in with can give rise to mental lethargy in the animal – feeding a dog once a day is a prime example.

If you've ever been on a long-haul flight, or even a medium distance air passage, can you recall what your thoughts were about the in-flight food? After spending the first hour or so getting comfortable, for the rest of the journey you're reliant on the flight crew to break up the monotony of the trip by bringing you food, drinks, or snacks, or watching a film/reading. Your environment, your stimulation is in THEIR hands, and there's not much you can do about it. You get to the point where you are so looking forward to the dried-out cheese sandwiches you are given to eat that it becomes a stimulating event in an otherwise boring passage of your life. You anticipate and imagine what the sandwich will taste like; you wonder what might be accompanying it – a little salad garnish, perhaps? It finally arrives – what a build-up! – but in a few short moments it's gone, and you're back to waiting and anticipating the next pass of the drinks trolley, or arrival of the in-flight magazine ...

Well, welcome to your dog's world!

Imagine waiting all day, looking forward to those fantastic few moments when your food is being prepared; you know it's coming, you can smell it, you can SEE it, even, this is your meal and no mistake, the anticipation is almost too much to bear. But, in a matter of seconds from when your bowl hits the floor, it's all over. Gone. Bored again. What can I do to bring a little excitement into my life? I know, I'll pull all the washing off the line outside. That looks like fun ...

Consider feeding your dog periodically throughout the day – or maybe giving two meals a day – without increasing the overall amount of food. This can provide a great deal of mental stimulation, and your dog will most likely love the idea!

When undesirable behaviour such as mischief-making, destruction of household items, signs of anxiety and even aggression manifest themselves, boredom born of a monotonous life can often be the root cause.

CLEVER DOG!

Introverted, stereotypical behaviour such as sucking, licking and biting the paws, wrists and flank areas can be another expression of boredom. But even if your dog is not exhibiting such signs, behavioural enrichment is essential to keep him happy and healthy.

And, of course, in 99.9 per cent of cases, your dog's environment and available behavioural enrichment is in your hands.

Stimulation: dog activity puzzles

Providing an enriching and stimulating environment is an important factor in having a healthy, well-adjusted pet. Feeding your pet at regular intervals by the usual means of his bowl will satisfy his nutritional needs, but does nothing to address psychological and emotional appetites.

Interactive feeding toys can be a great tool for breaking up feeding schedules and allowing your dog to work (scavenge) for his culinary reward.

Many zoos have included a variety of interactive feeding programmes in their animals' daily feeding schedule. Some of these programmes mimic foraging choices similar to those found in an animal's natural habitat, including hiding food, or hanging fruit from a tree.

If you've ever seen the late Steve Irwin, self-proclaimed 'crocodile hunter,' you will have witnessed how he risked life and limb to ensure his animals 'worked' for their meals. Standing by the edge of the pool, he agitated the water to simulate motion, and encourage the crocs to rise from the shallows to 'attack' the meat he proffered. Of course, he could just as easily have pitched the meat over the fence; the crocs would still get their meal, and, nutritionally, they'd be just as well off, though mentally they'd be missing out.

Other feeding programmes use unusual and challenging food items that provide mental stimulation, such as frozen mixtures of natural foods, or feeder puzzles. These programmes have added variety and enrichment to feeding time for zoo animals, and have helped combat depression and stereotypical behaviours such as pacing and swaying, common with animals in captivity.

How do interactive food puzzles work for mental stimulation?

• The food is distributed over a longer period of time. A food dispenser filled with chicken, lamb, liver, or even 30 pieces of kibble, lasts about half an hour, whereas the same amount of food would usually be eaten within thirty seconds to three minutes when the food is freely available in a bowl. The time spent foraging (looking for food) increases and the dog is mentally stimulated for longer

• The food is not permanently available but instead is available randomly. This unpredictability may raise a dog's vigilance, thus decreasing boredom levels

• Maintenance is easy and does not require any additional time. The food dispenser is given to the dog to play with and can be refilled any time

• Even small amounts of food delivered by the dispenser have a strong effect on the behaviour of the animals. This is important, because all enrichment activities related to food have to be incorporated into the feeding and training schedule; this is much easier when the amount of food needed for enrichment is small

• It's inexpensive. This is obviously important, because high additional costs are often used as an argument against behavioural enrichment. And you're not being asked to hang chicken carcasses from trees in the park, or drag rabbit skins along the ground for your dog to track and 'kill'

Interactive food puzzles. A rather simple, inexpensive piece of kit that can help to enrich your dog's foraging experience and thus reduce boredom and monotony.

Note that if you have more than one dog, it's a good idea to separate them when they're playing with food puzzles to avoid any potential conflict.

Scenting, tracking and foraging

As part of your dog's behavioural enrichment programme, scenting and tracking should be encouraged. If you don't have a garden where your dog can be let off the lead for a free run, maybe use a different room in your house to hide objects. Searching and tracking exercises are great activities for both under-stimulated and over-active dogs.

Playing hide and seek with your dog's favourite toy, or even small portions of food, is incredibly rewarding for him, and if you don't get that warm, glowing feeling watching his little tail wag feverishly when he makes 'the find' you should consider a trip to the doctor to see whether you actually posses a heartbeat!

The one stipulation with behavioural enrichment programmes is that they are individual exercises that should be done separately with each of your dogs if you have more than one.

Hide and seek – canine style

• Walk across a grassy area (preferably an enclosed garden), treading firmly to make imprints on the ground that your dog can scent
• Place a toy or interactive treat dispensing activity puzzle with some food somewhere discrete, but not impossible to locate
• Call your dog to you and encourage him to 'find.' He should be able to follow your trail to the hidden treat

'Hansel & Gretel' is another tracking game.
• Walk ten yards or so, drop your dog's favourite toy in the grass and conceal it slightly
• As you walk back, drop small amounts of food along the route
• Drop the last piece of food just in front of your dog and release him to follow the food trail
• When he gets to the end of the track, he has the additional surprise of finding his toy
• Scattering food around your garden and letting him find it is another great way to allow him to use his nose

For more great scenting games, see *Smellorama – nose games for dogs*, published by Hubble and Hattie.

Other games you can play with your dog include the 'predatory sequence game.' Dogs love to rip, tear and dissect! Wrap a piece of meat in a clean cloth, tying as many knots as you can so that your dog really has to work to get at the meat. And, of course, dogs love to dig. If you can allocate a part of your garden as a digging area, bury food, toys, bones and chews in the pit and let him find them.

Don't forget physical stimulation

Regular grooming each day not only keeps a dog in tiptop physical condition, it also promotes mutual trust and affection between animal and owner.

Make grooming a fun, regular activity and use the time to check your friend for lumps, bumps or abrasions. Loving your life with dogs means you want to spend as long together as possible, and regular touching and caressing of your dog's body, knowing it inside out, could save his life where cancerous lumps are concerned.

Remember, breed differences should be borne in mind when considering enrichment options. Your average Collie will enjoy robust, motion-based games; hounds will love scent activities, and Labradors will love anything where food is the ultimate reward!

Obviate undesirable behaviour by ignoring it and only rewarding desirable behaviour. It's a very, very simple rule to live by, and experts agree that positive reinforcement training will provide the most effective and long-lasting results.

The more your dog is actively engaged in positive play during sessions or training exercises, the less time he or she has to develop undesirable behaviour.

CLEVER DOG!

Lessons we can learn from dogs

Dogs are natural-born problem-solvers. They seem capable of understanding, very quickly, what is required to remedy a problematic situation in the fastest, most effective manner.

This is because most of the problems the dog has been required to address throughout the ages have been of life or death importance: finding food, evading predators, reproducing. Although the problems dogs face today aren't usually matters of life and death, the instinct remains.

Humans are lucky in that we don't have this type of problem every day of our lives. There's an 'at all costs' element to a dog's approach to solving a problem which is not usually the case with humankind. Whilst it's important not to 'sweat the small stuff' of life, it is important to resolve problems that worry us.

Dogs are hard-wired with the knowledge that if they don't resolve the issue, no-one else will, but are we, prehaps, too quick to try and pass on whatever is troubling us, rather than facing up to it and finding a solution? Or, worse, sticking our head in the sand and ignoring it?

Dogs will disregard all distractions when trying to get food out of a food puzzle. For those few minutes, that is their entire world. Adopting a similar attitude and approach may just make us a better problem-solver.

SIX

CANINE COPING MECHANISMS

How dogs cope with stress, conflict, illness and loss

"Dogs are the rockstars of the animal world. They live fast, die too young, but they have one hell of a time whilst they're here." – A Rayme

Dogs are just as inclined to experience stress, conflict, illness and loss as their owners. While your dog won't agonise over where his next meal will come from, or when you will next take him for a walk, he may well wrestle with his own fears and anxieties as much as any person.

These anxieties may be associated with a particular person, or a specific situation, such as a visit to the vet. Dogs may become anxious when their owners are away, when they hear certain noises such as thunder or fireworks, when a new pet joins the family, or an animal companion dies.

What causes our dogs to feel stressed?

Well, there are numerous, everyday things that can make our dogs feel stressed and anxious: physical pain, changes in the home environment, change in pack hierarchy, confinement, negative commands, separation, aggressive owners or aggressive animals, being hungry, and the inability to relieve himself when needed are all common causes of stress for your canine friend.

In order to alleviate the stress your dog is feeling, it's obviously important to determine what the trigger for it is. Dogs are much less prone to stress if they lead happy lives, which include plenty of exercise, grooming, and, of course, fresh water and food. Loud noises or a disruptive lifestyle can unsettle a dog. Eliminating these types of agitating circumstances will greatly improve your dog's attitude.

How to recognise stress in your dog

Stress is the body's response to physical or mental stimuli, and prepares the body to either fight or flee. It raises blood pressure, steps up heart rate, breathing and metabolism, and there is a marked increase in blood supply to the arms and legs. It is a physiological, genetically predetermined reaction over which the individual, whether a dog or a person, has no control.

When your dog is stressed, his body becomes chemically unbalanced. To deal with this imbalance, the body releases chemicals into the bloodstream in an attempt to rebalance itself. The reserve of these chemicals is limited, and can be dipped into only so many times before it runs dry and the body loses its ability to rebalance and achieve homeostasis. Prolonged periods of imbalance result in neurotic behaviour and the inability to function properly.

Your dog experiences stress during training, whether you are teaching him a new exercise or practising a familiar one. You should be able to recognize the signs of stress and what you can do to manage the stress your dog may experience. Only then can you prevent stress from adversely affecting your dog's performance during training.

Stress is characterized as 'positive' (manifesting itself in increased activity) and 'negative' (manifesting itself in decreased activity). Picture yourself returning home after a hard day at work. You are welcomed by a mess on your new, white rug. What is your response? Do you explode, scream at your dog, your children and then storm through the house slamming doors? Or do you look at the mess in horror, shake your head in resignation, feel drained of energy, ignore the dog and the children and then go to your room? In the first example, your body was energized by the chemicals released into the bloodstream; in the second example, your body was debilitated.

Dogs react in a similar manner when stress triggers either the fight or flight response. Positive stress manifests itself in hyperactivity, such as running around, bouncing up and down, or jumping on you, whining, barking, mouthing, getting in front of you or anticipating commands. You may think your dog is just being silly and tiresome, but for the dog, those are coping behaviours. Negative stress manifests itself by lethargy, such as freezing, slinking behind you, running away or responding slowly to a command. In new situations, he seems tired and wants to lie down, or sluggish and disinterested. These are not signs of relaxation, but the coping behaviour for negative stress.

Signs of either form of stress in dogs are muscle tremors, excessive panting or drooling, sweaty pads that leave tracks on dry, hard surfaces, dilated pupils and, in extreme cases, self-mutilation, urination or defecation, usually in the form of diarrhoea. Behaviours such as pushing into you or going in front of or behind you during distraction training are stress-related.

Tail chasing

Although it looks comical, dogs who chase their tails aren't necessarily as happy as they might appear. When they don't know how to deal with a situation, such as meeting a strange dog, chasing their tails acts as a distraction. It buys them time while they think about what they're going to do next.

This sounds like a silly way to cope with confusion, but people do similar things. It's called displacement behaviour and we do similar things like drum our fingers or jiggle our legs.

Too much stress can cause a dog to bite

Even the gentlest, most loving dog can be induced to bite. Dogs' teeth are important tools, and each is aware of their potential use as offensive or defensive weapons.

Every dog has a bite threshold (a point beyond which, if pushed, he will bite). Some thresholds are low; some are high. Aggression can be caused by stress, and each stress factor builds toward the point where a dog will bite. If a dog has a low bite threshold, it might not take too much stress for him to reach that point.

The four common canine stress triggers are:
• small children under the age of 4
• thunder
• men
• moderate to severe pain

The longer a dog's list of stress triggers is, the more likely he is to eventually bite, which is why early and ongoing socialisation is critically important.

How to help your stressed dog

• Try feeding your dog food that has wheat in it. Reports have shown that the by-products from digesting wheat can stimulate certain centres of your dog's brain, which can induce feelings of calm
• If your dog has a fear of thunder, fireworks, and other loud noises, desensitizing or habituating your pet is almost impossible. Instead, provide a place in your home away from windows and doors where he can wait out the storm, and keep a radio on to help muffle the sounds of the thunder or fireworks
• Crate training. A crate can be a 'safe house' for your dog when the world around him is changing.

Any time you travel, move, or leave your dog for short periods, put him in his crate with some comfortable bedding, a shirt or towel that has your scent on it, and his favourite toy. Accustom him to his crate slowly and carefully so that it becomes a place he can always count on to feel safe
• If your dog is likely to experience anxiety because a new animal has joined the family, introduce them to each other gradually, maybe on neutral ground away from the home initially
• If your dog seems anxious when you are away even for short periods of time, leave the television or a radio on as a way of providing some 'human' company
• If your dog becomes stressed because you are moving to a new home, take him there before you move in, if possible, and allow him to investigate the surroundings. Give your dog some treats or play with him in the new home so that he will associate something positive with the location
• If the appearance of a new baby or new spouse in your home is making your dog anxious, try to keep things as normal and routine as possible. Stick to your dog's regular schedule and give him plenty of attention

Dogs and conflict
Put two or more people under the same roof, and tensions occasionally flare up. If two people can't always keep the lid on emotional outbreaks, why should animals be expected to do so? Naturally, your animals will occasionally disagree or fall out.

Canine conflict aggression
Dominance is a very normal behaviour in dogs, and you will see varying degrees of it in every dog that you meet. Without a clear hierarchical pattern in a social group of dogs, there will be constant conflict over resources. A dominance pattern allows effective distribution of valued resources (food, shelter, territory, and access to females) among the members of the group. The higher ranking individuals dictate this distribution through the use of cues and signals that fall short of aggression. Only when individuals challenge the social order do aggressive displays result.

How to handle canine conflict
If your canine housemates get into a squabble, try some of these methods to separate the animals without risking life or limb:
• Proper introductions and training will go a long way toward preventing fights
• Water is one of the most effective and harmless ways to separate two dogs that have locked horns. If the dogs are outside, squirt them with a hose. Keep a well rinsed out washing-up bottle filled with water which can be used indoors to deliver a timely squirt of water
• Give your dogs their own food dishes, beds, and toys to discourage competition that may cause them to fight
• Forming a hierarchy is natural to dogs; respect the hierarchy by feeding the alpha (dominant) dog first, and by not giving dogs lower in rank special attention or favours
• Try manoeuvring the dogs into a doorway and separate them with the door
• On no account, put your hands near their heads or teeth, by trying to grab their collars, for example. Without doubt, you will get bitten if you do so
• Lastly, adopting or purchasing a dog known to get along with another species will save you time and energy helping each adjust to the other. A dog that has had dealings with cats and other dogs is more likely to get along with a newcomer (assuming the exposure was a positive one). If you got your dog from a shelter, ask the staff if they know anything about his history, and whether he came from a home that had both dogs and cats. If you purchase your dog from a breeder, ask about what other animals the breeder has in his household

How dogs cope with illness
No matter what the illness or injury, once a dog or cat becomes familiar with any limitations that this might pose, he generally adjusts quite readily to living within his newly defined parameters.

CLEVER DOG!

And just as often, he astounds us by his refusal to be limited, to act the invalid, or to be waited on by doting and often guilt-ridden owners. He does the best he can, often far surpassing what is expected of him by his humans, and occasionally returning so nearly to his former self that the uninformed will barely notice his affliction.

Dogs learn to compensate for missing limbs and faculties by utilizing other resources at their disposal. Thus, the three-legged animal soon builds up muscles in his remaining limbs to compensate for the missing one.

Epilepsy

Idiopathic epilepsy is quite a common occurrence in domestic animals, frequently seen in the German Shepherd, Poodle, and many of the smaller, more excitable breeds. However, any animal, mixed or purebreed, can be born with epilepsy, or become epileptic.

Convulsions can occur as infrequently as two or three times a year, or as frequently as several times a day; as mild as a slight twitching, or as severe as violent running fits. Many vets do not begin treating epilepsy until the seizures increase in intensity, duration, and/or frequency.

The epileptic animal may experience convulsions during periods of excitement when the general household tenor is at high pitch, or when subjected to stress caused by travelling, or exposure to strange people or places. In these instances, the animal may be sedated or confined to an area where a quiet, relaxed atmosphere prevails.

Modern medication, given orally on a daily basis, is generally all that is necessary to completely eliminate (or greatly limit) the severity and frequency of epileptic seizures, thus allowing your dog to live a relatively normal life.

When convulsions do occur, first aid consists primarily of preventing the animal from injuring himself during mad dashes around the house, thrashing about on the floor, or other spastic behaviour.

Remember, though, that your pet is not in control of his actions while in the throes of a convulsion, so keep your hands away from his mouth and teeth. Contact your vet should the convulsion last more than a couple of minutes, or seem unduly violent. In the aftermath of a convulsion, ensure your dog has access to water, as well as an area to use as a toilet. If he needs to go outside to do this, go with him to ensure he doesn't fall or hurt himself. Encourage your dog to lay in his basket or somewhere he finds comfortable, and remove all noise and stimulations, such as the radio/TV, etc, so that he can relax in peace.

Diabetes and daily treatment

Treatment usually calls for stringent dietary control, a daily urine sample check for glucose, and, of course, daily (or sometimes twice-daily) injections of insulin. Animals, luckily, have a vast subcutaneous layer into which the insulin may be injected, thereby causing far less irritation than for similarly afflicted humans. It's also possible that diabetes in some animals can be controlled by diet alone, but this will need to be decided by your vet.

After an initial period of adjustment for both owner and animal, the routine of caring for a diabetic becomes second nature.

Paralysis

Cherie Kendal was just a normal, devoted dog owner, until her dog was paralysed in an accident which meant she had to radically rethink how she went about her life. Since the accident, Cherie has focused on exploring animal disability and care, whether caused by trauma or illness. Paralysis entails a considerable adjustment for a person, but is surprisingly easy for a dog. She explains:

"Paralysis in a dog is often a stunning blow to owners, though, generally, he will be in little or no discomfort, apart from some confusion, which is only natural.

"Frequently, bowel and bladder control is impaired or temporarily lost, in which case, hospitalisation will be necessary to express the animal manually, and help him adjust until he is accustomed to his condition. When control returns to the sphincter muscles, the animal may be discharged with a 'doxie cart' which will allow him to remain mobile while waiting to regain control of hindleg movement.

"This gradual improvement can take months, but when he is able to use his hindlegs again, a prophylactic surgical procedure known as fenestration may be undertaken to prevent paralysis from reoccurring.

"Handling the paralysed animal requires little effort on your part; placing him in the cart several times a day for exercise periods, and to allow him to relieve himself, is a simple task.

"It is always essential to provide adequate support when carrying such animals by placing one arm between the front legs and under the sternum, and the other between the hindlegs and under the abdomen. This will prevent further injury to delicate spinal tissue.

"Once properly fitted with a doxie cart, the paralysed animal soon becomes quite self-sufficient, learning to back up, navigate around obstacles, and even lie down. Of course, he must not be left unsupervised while sporting his 'wheels,' in case of accident.

"Front leg paralysis can be mitigated using mechanics in the same way as hind leg paralysis, although paralysis in the area around the front legs does pose other general problems which are best discussed with a vet."

How dogs adapt to an illness or disability

Broken bones and minor lacerations are undoubtedly as common with animals as they are with children. The handicapped dog quite frequently suffers little or no permanent psychological impairment, and promptly learns to live with, or around, his disability, no matter how great.

Trauma is trauma, whether experienced by a human or an animal, but certain factors exist for a dog which give him an enormous advantage in coping with traumatic injury or serious disease. After the necessary initial recuperative period following major surgery, say, your companion should once again be capable of functioning in a manner not very different to his usual self.

The dog who has lost his sight, for example, learns to use his senses of smell and hearing to a greater degree, providing necessary clues to his environment. The animal whose hearing has diminished, or is non-existent, employs other faculties to make up for this lack.

Many diseases, such as diabetes and idiopathic epilepsy, require only minimal daily medication and care to assure a happy, healthy existence. Even severe problems such as hind-end paralysis, commonly associated with intervertebral disc disease, can be managed at home once the animal's condition has been treated and stabilized.

There is no doubt that the owner of a handicapped dog must make some adjustments, more, perhaps, than will be required of the animal himself. Your pet feels no embarrassment about his condition, nor does he consider himself any less valuable as your companion and friend – but you might.

Please try and keep in mind that your dog will not be resentful or hostile about his infirmity, nor will he spend anxious hours worrying what his friends or family will think of him now that he's disabled. He will accept his somewhat modified life with equanimity and good nature, and will make every effort to return to a relatively normal existence as quickly as possible.

Disabled or loss of vision

Losing an eye can be a problem for anyone, animal or human. However, since dogs have far more sophisticated senses of smell and hearing, they have a distinct advantage over their human counterparts in this respect.

Fellow family members of an animal who has lost an eye should learn to approach him from his sighted side, to avoid startling him if he is not otherwise aware of their presence.

For the animal who has lost vision in both eyes, whether due to old age, disease, or injury, management is slightly more complicated, but by no means impossible. Take care to always alert him to your presence by calling his name so as not to startle him.

CLEVER DOG!

The position of furniture should, if possible, be left unchanged – or if a rearrangement is necessary, gently introduce your dog to the new arrangement, and familiarize him with it before he is left unsupervised in the area. You may want to make stairs and steps off limits to your dog, although this isn't always necessary as many blind dogs cope with these admirably. Baby gates (available at department stores or pet shops) are an easy way to restrict particular areas of the home without limiting human access to such places.

Allow your dog to sniff his food before it is placed in front of him (you may have to guide him to the bowl initially). Routine is extremely important to managing sightless animals, many of whom are older pets with multiple problems. If loss of vision has been a gradual process, your dog may well remember where food, water, furniture, a favourite bed, and the door to outside are located.

At all times, give your dog verbal encouragement and reassurance to let him know that all is well, and that the world he once saw is still there.

An animal with seriously impaired/no vision should not be left unsupervised in unfamiliar places or outside where there are obvious dangers. It's far better to confine such a pet to a small, safe room than to allow him to wander through unfamiliar areas. (For more information see *Further reading.*)

Loss or impairment of hearing

That your dog may have impaired hearing – or has gone copletely deaf – is not always obvious, except when total deafness is congenital, as in some Dalmatians.

Owners of deaf animals, or those that are hard of hearing, learn to use vibration to communicate with their pets, stamping on the floor to alert the animal of their presence, for example. Training a deaf dog entails using hand signals and lead corrections. Deaf animals, obviously, must not be allowed to run free, since they are at a distinct disadvantage in an unprotected and unrestricted environment. (For more information see *Further reading.*)

How dogs cope with limb loss

Veterinary science is moving at quite an exciting rate, and dogs are living longer and surviving once fatal diseases because of that. Researchers and businesses which support the veterinary community are playing their part in helping our animals live a healthier, longer life, too, but one incidence that cannot be medicated against or legislated for is trauma.

Being hit by a car, a household accident, falling or being fallen on can all result in the loss of a limb for a dog, and certain illnesses can lead to this sad conclusion, too. Could it be that the secret to the dog's success as a natural born survivor lies in his inability to conceptualise the impact of disability?

And we should not lose sight of people like Dr Glyn Heath, who is doing extraordinary work to ensure that life is as comfortable and mobile for those dogs that do undergo amputations. Here he explains how he goes about providing prosthetic limbs for canine amputees.

"In order for a dog to be able to walk effectively, it is critical that all four limbs are intact and that they are functioning properly. However, with many dogs this is not the case, as, for example, after limb amputation. Limb amputation is a relatively common surgical procedure in dogs and the most common reason for amputation is due to trauma, where the injured limb cannot be saved. However, there are other reasons for amputation and these include osteosarcoma or soft tissue sarcoma.

"In many cases, dogs can cope well when walking on three legs; however, sustained mobility may be a problem in larger or heavier breeds, and prosthetic replacement can improve their mobility. Prosthetic restoration does offer a viable alternative when a significant proportion of the lower limb can be saved, especially if the hock or wrist remains functional. There are also dogs that suffer paralysis for a variety of reasons. Partial paralysis of a limb creates considerable problems when walking as the limb does not function optimally and may well be insensate, the result being that the dog will damage its limb during walking. It is possible to produce an exoskeletal device (orthosis) that supports and protects the paralysed limb, without compromising other non-affected joints.

58

"With regard to dogs who have partial paralysis of one or more of their limbs, the situation very much depends on the severity of the paralysis and the joint movements that are affected. In many cases, the supporting role offered by orthoses can allow the limb to be functional, and by ensuring that the paralysed limb is held in the correct position, we can further ensure that the limb can support the body weight during stance and strike the ground during walking at the correct angle and with sufficient stability.

"Myself and Mr Geof Riley at the Directorate of Prosthetics and Orthotics at the University of Salford are involved in the design and fitting of prostheses and orthoses, utilising biomechanical principles similar to those applied when working with human patients. This work has been done in conjunction with veterinary surgeons, who have felt that the rehabilitation of selected patients may well be enhanced. To date, about fifteen dogs have been treated. Most of the patients have been fitted with orthoses for correcting gait, and many of the dogs treated have had notable improvement in their mobility.

"Where amputation is deemed necessary, it is usually carried out though the pelvic or pectoral girdle. Such an amputation is not ideal for prosthetic restoration as there is no residual limb remaining that can be inserted into the prosthesis. As such, the amputation should, where practically possible, preserve as much limb length as is possible with a view to retaining the functionality of all joints above the level of the damage. Limb reconstruction is a possible alternative, but is a relatively expensive procedure, and there is the added problem of bone rejection and unsuccessful grafting. Treating infection of the grafted bone is particularly difficult as the grafted bone is reliant on a reduced blood supply, and is thus more difficult to treat with antibiotics.

"The procedure by which a prosthesis or orthosis is fitted is remarkably similar. In the case of a prosthesis, a cast has to be taken of the residual limb, which has to include the area of the stump, and the area of suspension as far up as to the place of suspension. Taking the example of a dog that had an amputation halfway across the hind foot, the cast would extend to just below the stifle joint. The cast would have the position of the hock joint marked as well as areas susceptible to pressure, such as the end of the stump. The cast would then be filled with plaster of Paris, making a positive mould, and this mould will have plaster added and taken away from specific areas depending as to whether they were pressure-tolerant or pressure-intolerant. It is from this positive mould that a socket is made from polyester resin cotton and fabric with fibreglass reinforcement. Once the socket is made, the extension section will be added so that the prosthetic leg is the same length as the anatomical leg.

"Orthoses are made in a similar way whereby a plaster of Paris cast is made over the affected leg and a mould taken from the cast. An orthosis may be made either from plastic or a polyester resin composite similar to that of the prosthesis. In this case the orthosis is for a dog with a ruptured Achilles tendon. The animal could not fully extend its leg but the orthosis supported the leg such that it could walk on it, albeit with a stooped gait.

"Dogs appear to adapt very quickly to their prosthesis/orthosis, and in some cases they are mobilising reasonably well after only a few minutes. However, their ability to use such appliances depends very much on how well they fit, level of comfort, and their alignment when walking. It is these areas that require specialist attention, which is made more difficult because dogs are unable to communicate their problems as easily as humans. The most critical problems that can arise are that sections of the dog's skin may be exposed to high pressures, and rubbing of the device against the fur/skin. If such problems are not resolved during the fitting process, then pressure sores and loss of fur may result. However, if prostheses/orthoses are comfortable and correctly aligned, they can provide enhanced mobility for the patient. The benefits to the owner are greater because their pet can move around without excessive effort."

Loss and grieving

Like humans, dogs grieve when they lose a loved one, be they human or animal family members. For dogs, however, the process must be so much harder: unlike us, they can't pick up the phone and

CLEVER DOG!

share their pain with a friend, or get stuck into a bottle and blot out the memories of yet another day filled with heartache. For a dog, grieving is a lonely and desolate experience, and should be taken seriously ...

Greyfriars Churchyard, Edinburgh, 1858

A dark, grim afternoon encompassed the mourners at the funeral of a Mr John Gray. At the end of the service, one by one, his family and friends drifted away – their last heartfelt words echoing in the cold November air. But one mourner was not to be moved. Greyfriars Bobby, a Skye Terrier, owned and adored by John Gray, curled himself on his master's grave and closed his eyes. He was not to leave for fourteen years.

This true story is the tale of the most famous dog-grieving episode of all time. Bobby slept at his deceased owner's grave every night without fail for fourteen years, until his own death in 1872. In honour of Bobby's amazing devotion a statue and water fountain were erected a year later in the churchyard.

Jacqueline Pritchard is an animal behaviourist and she believes it is time we took dog grief seriously.

"There is no doubt that dogs grieve just like we humans do," says Mrs Pritchard. "I remember the worse case I had to treat was a dog who had lost an owner and just completely went into himself. He simply went under his dead owner's bed and stayed there. There was a real chance that the dog would have to be put down, just to put him out of his misery. In the end, the situation was resolved by putting the dog on a course of anti-depressants."

One of the big problems with dog grieving is that often, naturally enough, others will also be grieving. It's a fact that, when we lose a loved one, we focus first on our children, our siblings, or our parents, which means that our dog may not just receive less attention at a difficult time, but also receive no comfort or understanding in his grieving. This is when friends can help by taking your dog out for a walk with theirs, hopefully lifting his spirits a little in the process.

But what of dogs who lose their only carer, which is true of many dogs who have elderly owners, just like Greyfriars Bobby. Jacqueline Pritchard says that, surprisingly, these dogs actually often recover more quickly.

"Dogs who are left ownerless are taken in as rescue dogs and rehomed, which means a whole new environment straight away. Although that may sound very harsh, it is, in fact, the ideal way for a dog to move on and get over his loss. A new environment means no reminders of the owner who has left them behind."

It's important to remember, then, that dogs – like humans – also grieve, for their human companions and their animal friends. As with us, time, love and understanding are what a dog needs to get over the loss. Being on the lookout for early signs of depression can help make your pet's grieving process as short and pain-free as possible. (For more information see page 95.)

Dogs and pessimism

An RSPCA-funded study by the University of Bristol has revealed new insight into the minds of dogs, discovering that those which are anxious when left alone also tend to display pessimistic-like behaviour. The study allows important comprehension of canine emotion, and enhances our understanding of why behavioural responses to separation occur. Study leader Professor Mike Mendl, Head of the Animal Welfare and Behaviour research group at Bristol University's School of Clinical Veterinary Science, said:

"We all have a tendency to think that our pets and other animals experience emotions similar to our own, but we have no way of knowing directly because emotions are essentially private. However, we can use findings from human psychology research to develop new ways of measuring animal emotion.

"We know that people's emotional states affect their judgement, and that happy people are more likely to judge an ambiguous situation positively. What our study has shown is that this applies similarly to dogs – that a 'glass-half-full' dog is less likely to be anxious when left alone than one with a more 'pessimistic' nature."

In order to study 'pessimistic' or 'optimistic' attitudes, dogs at two UK animal rehoming centres were trained that when a bowl was placed at one location in a room (the 'positive' position) it would contain food, but when placed at another location (the 'negative' position) it would be empty. The bowl was then placed at different locations between the positive and negative positions. Professor Mendl explained: "Dogs that ran fast to these ambiguous locations, as if expecting the positive food reward, were classed as making relatively 'optimistic' decisions. Interestingly, these dogs tended to be the ones who also showed least anxiety-like behaviour when left alone for a short time. Around half of all dogs in the UK may at some point perform separation-related behaviours – toileting, barking and destroying objects around the home – when they're apart from their owners. Our study suggests that dogs showing these types of behaviour also appear to make more pessimistic judgements generally."

Dr Samantha Gaines, Deputy Head of the Companion Animals Department from the RSPCA, said: "Many dogs are relinquished each year because they show separation-related behaviour. Some owners think that dogs showing anxious behaviour in response to separation are fine, and do not seek treatment for their pets. This research suggests that at least some of these dogs may have underlying negative emotional states, and owners are encouraged to seek treatment to enhance the welfare of their dogs, and minimise the need to relinquish their pet. Some dogs may also be more prone to develop these behaviours, and should be rehomed with appropriate owners."

Lessons we can learn from dogs

Dogs don't fret about their problems or disabilities simply because they don't understand them in the way that we do. We DO have a concept of what being disabled can mean for us, especially if we have been forced to make the transition from able-bodied. But that doesn't mean we can't take inspiration from a dog's approach, and appreciate the value of that old – some may say cliched – virtue of a positive mental attitude. Dogs, in some instances, seem able to adjust to serious physical disability in, as we learned in this chapter, 'a matter of minutes.'

Natural born survivors, perhaps they don't have the indulgence of retrospective thought; they don't spend time remembering how great things used to be. They live in the moment and approach each new challenge with a simplicity that is remarkable and commendable. They do appear to grieve. They certainly can become stressed, but do you know of a dog that actually died an early death as a result of stress and strain?

The dog's renowned abilities as a survivor, the ultimate coping mechanism we are all blessed with, is probably best characterised by his overwhelming insistence on living for today. Similar to stories of children being better equipped, mentally, to survive in the rubble of earthquakes, dogs, too, are able to switch their immediate focus to rapidly identifying and attempting to resolve problems one moment at a time. They don't take the weight of the world on their shoulders, but certainly have an enviable zest for life that keeps them in a constantly positive frame of mind. When it comes to coping with adversity or conflict, humans are too proud. Dogs aren't burdened with a sense of pride. That's why a three-legged dog doesn't think twice about chasing a four-legged dog around the park. Dogs don't fight for the sake of pride; they'll only ever fight as a last resort. They know that life is too precious to go into battle over something trivial, but people do it all the time. We argue with our neighbour over parking spaces when there's a space down the road. We flash our lights at other drivers because we feel we have right of way. We do things because our pride tells us to.

When dogs come into conflict, both animals will do everything they can to contain a situation before things 'get silly.' We waste our time battling for things 'on principle,' 'standing our ground' to prove we are right. If we lost our petty pride and concentrated on doing things that improve our lot, and the lot of those close to us, we'd get more done and waste less time. The world would be a better place as a result, as well as a lot more fun!

In this respect, we can learn an awful lot from our peacemaking canine friends ...

SEVEN

THE CANINE MAGICIAN

Dog tricks with a difference

"The dog has the answers to questions we have yet to even ask of him." – Charles Vincent

In this chapter, we're going to take a look at the amazing canine abilities that can help us, deceive us, and astound us, as well as consider what other, so far hidden talents we may yet have to discover in our faithful friend.

Ever made a 'snap decision?' Purchased on impulse, taken an unfamiliar route on a whim? Then you've used rapid cognition, or, as Malcolm Gladwell author of *Blink*, describes it "thin slicing;" the decision you make within a couple of seconds of a new situation presenting itself.

Taking an instant dislike to someone – or the opposite, love at first sight – is another example of how humans use rapid cognition, and yet as a species we have learnt to second-guess these instincts. As information has become power, we have discarded the idea that our very own nature and biological programming has anything to tell us. Our 'gut instinct' has to be justified by tangible facts, and, in an age of unprecedented litigation, proof that the snap decision we make is the right one.

Take your local accident and emergency department, for example. If you've visited recently (I hope you're feeling much better now ...) you will have first reported to reception, which is then followed by triage to assess your importance in terms of medical emergency, depending on a set of questions. You may then have an interminable wait, quietly spouting blood, whilst your details are entered in a computer. Eventually, when you have all but lost the will to live, you will be seen by the most junior physician available to assess whether you should continue in the medical chain and in which direction if so. Tests are run and X-rays taken. You know you have a broken leg, the receptionist knew you had a broken leg, the triage nurse knew you had a broken leg, and the doctor now knows you have a broken leg. However, according to the rules of the system, they are powerless to act upon this fact until it has been irrefutably *proven* that you have a broken leg!

Now, I appreciate that this is a somewhat tongue-in-cheek over-simplification of our wonderful health service. But when this scenario is compared to one where time is of the essence, such as a war zone when a single second can mean life or death, and rapid responses and diagnoses are relied upon to ensure not just the safety of one patient, but that of whole platoons of soldiers, it does seem rather unnecessary, and not just a little ridiculous, even. In life-or-death situations, we trust the instincts of our soldiers completely (although afterwards even this can come under scrutiny if split-second decisions turn out to be erroneous; hindsight, eh, it sure is a wonderful thing ...).

Conversely, canines still live very much moment to moment. Not bound by the obvious and unspoken rules of human society, they scratch when they want to, sleep when they need to, and fight when they absolutely must. Whilst dogs have a very complex system for interacting with each other, which is universal (a dog from China will know exactly what a dog from Russia is conveying through his body language), we have managed quite enthusiastically to evolve to a state where we have little or no idea what dogs are trying to tell us. And it's not for want of trying on the part of the dog that this has happened; he has even evolved his own vocal language to mimic our own patterns of communication.

The domestic dog not only barks, he has a range of barks. From the high pitched, rapid, fear bark, or 'farm bark,' right through to the low, single 'play bark,' all of which have evolved in a rather endearing way to allow dogs to communicate in a way we humans find easy to understand.

At the risk of anthropomorphism, the dog in the modern world is living the role of a foreign exchange student. He has an understanding that certain things he does are either correct or incorrect, and appreciates many of the major signals we convey to him (language not being one of them). We've already given dogs all the information they need to complete a command before we vocalise it through our body language or 'tells' – subtle and often unconscious changes in demeanour that give us away. Yet the vagaries of society and custom, and subtleties of communication, are lost on him.

For example, teaching a dog to bark when the doorbell rings is a classic 'trick' that many owners end up regretting. We have 'taught' in our Pavlovian way, an animal to respond to a stimulus with a volley of barks to deter intruders and double-glazing salesmen. The dog thinks this is the correct and desired response every time the sound occurs.

The owner decides to have a party, or the doorbell breaks and won't stop ringing every few seconds. The dog is doing exactly as the owner has taught him. He is barking at the noise, and making a very fine job of it too. Yet now he is in trouble; he knows something's not right as his human is now responding angrily every time he does his job.

He has no idea that there are times when he is not required to do his stuff, as his owner has not taught him when to stop. As far as the dog is concerned, it could even be that, when his owner shouts at him for barking, he is joining in!

Sadly, this situation all too often ends with the dog being strongly 'corrected' by a human at the peak of frustration, sending horribly confusing signals to the poor creature, who may well go into meltdown.

Dogs are versatile learners who prove time after time that they are capable of adapting to our rapidly evolving society.

RSPCA staff in Oldham, Greater Manchester, taught a Polish dog to understand English. Cent, the Border Collie, was taken to the RSPCA when his Polish owners could no longer look after him.

Staff initially thought that he may be deaf but, after testing his hearing, they realised there was a language barrier.

Animal care assistant Luke Johnson said: "At first, we were baffled because Cent couldn't understand what anyone was saying to him. It was only a few days later it dawned on us that he must be used to hearing commands in Polish."

Staff at the centre searched on the internet for some basic translations, and also asked the former owner's family for assistance. Four months later, after using a reward-based training method, Cent is now bilingual, and responds to commands in both English and Polish.

Luke added: "As soon as we started using the Polish commands, Cent responded instantly, which was great. He really has come on in leaps and bounds since.

"It's not something you would really think about but it makes sense that, because our society is multicultural, so, too, are its animals."

Deceptive body language

Humans sometimes mistake certain canine actions as signs of 'guilt,' or other human emotions.

CLEVER DOG!

One of the most widely misinterpreted messages that a dog attempts to communicate to humans is the act of submission. Aggressive dogs are far less likely to attack a submissive animal, and we should expect our dogs to try and appease us when they sense that we are tense. Some dogs are so in tune with their humans that they appear to have a sixth sense, knowing when their owner is becoming angry, and exhibiting signs of submission before we have even acknowledged our state of mind ourselves. Here's a scenario that many dog owners will be familiar with.

You return home after a few hours to find your pet crouching slightly, tail tucked but wagging slightly, ears back, avoiding eye contact. The last remnants of a loaf of bread that was previously on the kitchen worktop is now in the dog's basket, and your dog's belly is very round indeed.

One interpretation of the dog's body language is that he is demonstrating the human emotion of guilt. The belief is that your dog knows he has done something wrong because he has 'stolen' the bread. He should be ashamed of himself and, seemingly, he is.

In reality, the dog has been using his inherent survival instincts. The pack leader has refused and walked away from some surplus resource (the bread) to which he, as the lower ranking pack member, can now enjoy free access. Your dog feels joyful excitement when you return and he enthusiastically, yet respectfully, greets you and welcomes you back into the pack.

You are taller than your dog, standing over him and, as he sees it, dominating him. You are also staring intently, another dominant gesture, and your body language is tense because you are now minus a loaf of bread and will have to replace it. You even smell different as your chemical balance alters slightly, due to your mood. And you will have communicated all of this to your dog within just a few seconds.

Your dog senses that something is wrong, and may now react by grovelling and frantically wagging and licking in order to avoid a reprimand or, worse, an attack. Some very anxious dogs may even urinate a little at this point to make sure you really get the message: "I am not a threat, be gentle with me."

You are a caring owner and would never hurt your dog, but he is living in the moment and has forgotten about the bread he ate an hour ago. He doesn't relate your dominance cues to his earlier 'mistake,' and his actions are merely his attempt to appease you.

Confusingly, a dog does sometimes seem to be trying to communicate in a similar way to us humans. Owners often describe dogs that 'cry' when left alone, and those that 'grin' at them. When we accept these messages for what they are – vocalisations to contact absent pack members, a submissive grin – the dog's response is understood and we can react more appropriately.

Mistakes people make in reading dog intent: are humans fully in tune with dog body language?

Television clips shows have much to answer for in our continuing amusement by obsessive and potentially dangerous behaviour manifestations. One only has to watch a popular TV show on a Saturday evening to see animals doing 'funny' things, such as obsessive tail-chasing, dragging small children around by their clothing, barking furiously at other animals, and in one unforgettable clip, actually attacking a car. This was a Karabash (Anatolian shepherd) in full aggressive mode protecting his flock of sheep.

Confusing obsessive behaviour with the dog being a 'joker' is something we have always found amusing, from the nipping lap-dogs of noble ladies who were erroneously thought to be 'thinking they are big dogs,' to the excitable Staffordshire Bull Terrier dragging his young owner across busy roads because 'he wants to say hello' to a cheery squirrel.

Recognising and responding appropriately to dog behaviour is something we, as dog owners, simply must improve on. By the time most owners recognise a potentially undesirable behaviour, it's too late. Dogs make their decisions in a split second, and only close reading and understanding of body language will improve the dog's lot in our society, where many of us live packed together, with dogs having to meet and deal with 'invaders' of their territory on a daily basis. Your dog is always giving you signals about how he is processing stimulus, and gathered here are some examples given to me by various owners I have worked with:

• *He just wants to say 'Hi'*
We've all met this fella. He's usually driven to the park, or drags his owner to the gate, where he's released and left to get on with it. Ignoring all canine protocol (polite dogs move slowly toward each other in a long curve), he runs full pelt in a straight line towards you, your kids, and your dog, his owner merrily bellowing from some distance away "It's OK! He's friendly!" just as the canine catastrophe leaps all over you and your animal.

Are you grimacing yet? Then, yes, you've met him!

Humans greet face-to-face; it's one of the adaptations to walking on our hind legs. Dogs acknowledge from some distance, and, using their tails, ears and general posture, will inform another dog if it is acceptable to approach them.

This guy hasn't been taught good dog manners, and ignores all protocol by running over and usually leaping on the (by now pretty annoyed) other dog. If you're lucky, your dog will be confused but tolerant. If you own a dog that does not appreciate this rude invasion and corrects in a very canine way, usually with a tightening of the body, followed by a growl (and if the other dog persists, possibly a scuffle), then be prepared, because the owner of the invader will probably decide that your dog is aggressive!

Humanising this dog does nothing but alienate him from healthy interaction with his own kind; he is unwelcome wherever he goes, and is an annoyance to human and canine alike. An excitable child can be informed that his actions are not acceptable through discussion; an excitable canine needs to be socialised and shown how to behave around his own species in a controlled environment, such as a reputable training class. In his favour, he will probably be so pleased to have this structure in his life, that you will see a rapid improvement!

• *She doesn't like men/vacuum cleaners/a broom/she must have been abused*
A common misconception is often bandied about with rescue dogs. Should they show a fear reaction to an object that, in theory, could have been used to abuse them in the past, and humans make the wrong assumption, a mildly apprehensive dog is encouraged to show fear of household objects. Your dog is asking you to show her how to cope with these sudden uncertainties via her body language, and soothing her only reinforces the notion that there is something to be afraid of.

• *He's not aggressive, he's only jumping up because he likes you!*
I could write a chapter on this one alone, and it does deserve further investigation. Suffice to say, if a dog is leaping at your face, invading your space, and putting his head at a level with yours, he's not aggressive. Yet ...

• *Selective hearing*
Your dog has wonderful hearing; in fact, even if you were to give all of your commands in the merest breath of a whisper, your dog is more than capable of responding. We rely too heavily on vocal commands with dogs, when what they understand better than we do ourselves is our intent. The dog running wild in the park can hear you calling her just fine, but would *you* come back to a frustrated and irritable person who sounds like they might be going to punish you? Hmmm, didn't think so. Think in her language, and use an inviting body stance and happy voice; you'll find that her hearing's just fine after all!

Better understanding the dog's value system
Do humans have a sound understanding of the canine chain of importance? For example, many children get bitten by dogs simply because adults have not recognised certain signals: do humans comprehend the fact that dogs do not understand human laws, and have their very own unique canine 'rules?'

The number of children bitten because they thought a snarling dog was 'smiling' at them increases every year. In general, dogs relate much better to children, as they are impulsive and react

CLEVER DOG!

very naturally to situations that an adult would second-guess. But dogs will treat these little people very much the same way that they would treat another dog if left to their own devices

Cartoons and feature films have humanised the dog so realistically that children are struggling to understand that their canine companion is anything other than an animated cuddly toy, with human emotions and fully understanding of their intent. Kids, though, do hold a key to a successful relationship with our canines – they're still curious, trusting, still full of joy, and still ready to learn – characteristics our dogs share. Watching a child and their trained, calm pet is one of life's great pleasures, and introducing a child to a creature that relies upon them for happiness gives them a great chance to learn empathy for others who are different from themselves: a wonderful life lesson.

In turn, the dog shows the child new ways to play, makes them feel safe and protected, is an ear to listen when they need it, and teaches them about trust and universal empathy.

In Argentina, a dog named China is hitting the headlines for finding and carrying a newborn premature human to her litter of puppies after she discovered it in a field on a cold night. The mother and her litter of pups kept the baby warm throughout the night, almost certainly saving the infant's life.

Kids take these lessons with them through life. The following is one of my favourite examples of rapid cognition in humans, and a teenager who has obviously been taught to value canine life (source *Mirror* online UK).

Bradley Thorpe, 17, risked his life to save year-old Husky Ben after he slipped his lead and jumped into the River Yare at Great Yarmouth, Norfolk. The student grabbed the dog and clung to a ladder in strong currents, eventually returning Ben to his owner, Desmond Noone, 50.

Bradley, of Yarmouth, said afterwards: "I just jumped in without thinking. The current was very strong. If I had let go of the ladder I would have ended up in Felixstowe – or even drowned."

Mr Noone, of Yarmouth, said: "I am so grateful." Coastguard Matthew Thornill said: "Bradley should be extremely proud – he was very brave."

Could a dog's vision be responsble for mistaken behavioural problems?

Most people are aware that their dog's senses of hearing and smell are keener than those of humans. Not as commonly understood, though, is the dog's deficiencies in some sensory processes, and whether or not this leads to problems often depends on the way people interact with them. The following helps explain many canine behaviours and reactions that owners don't understand.

Dogs don't recognize details within an outline, such as noses or eyes on a human face, but are fairly good at perceiving outlines. In a Pavlovian experiment, dogs were trained to discriminate between perfect circles and egg-shaped outlines. They performed nicely, but, when the ellipse was gradually rounded until it was $8/9$ths of a circle, the dogs failed to recognize the difference, a distinction most people can readily make. If repeatedly asked to do this, the dog lost all of its previously learned responses, even the big differences between circles and ellipses. Many dogs became neurotic about the whole thing, and had to be retired to kennels for a rehabilitation programme of rest.

In real life, an owner's hands usually signal positive treatment, such as petting and giving food. When the same hands inflict punishment or pain, the dog usually displays momentarily ambivalent behaviour.

Most owners are not aware that their puppy's vision does not reach maturity until about four months of age. Until then, objects are seen in various degrees of fuzziness, which makes visual identification of objects and individuals difficult, causing some pups to bark or growl at family members. If punished when this happens, the pups become confused and the seeds are sewn for problems such as submissive wetting and biting.

Scenting ability – superhero dogs and the nose that powers them!

The amazing power of a dog's ability to pick up on a scent has been demonstrated as a key service to humans by several dog breeds. For example, the German Shepherd is specifically suited for police work, and is one of many breeds that breeders have taken a keen interest in developing for their scenting skills, particularly with regard to hunting.

Consider also the bird dog, bred for his ability to track airborne scents.

Scent hounds pick up on the scent of their prey and hold onto it until their owner/hunter arrives and assists with the final kill. Such scent hounds include the Otterhound, Basenji, Dachshund, and Beagle, as well as many others.

During the 17th and 18th centuries, when the horrible and cruel practice of badger digging was rife, the Dachshund was used to sniff out badger setts in Germany. The dog's legs were short and perfectly sized to enable him to furrow into badger holes in order to grasp and hold the hapless creature until the dog's master would pull him from the sett by his tail, dragging the badger with him.

Saving lives through scent tracking

Snowy mountains and high altitudes don't faze the amazing St Bernard, whose unique service stretches back many years. These good-natured, impressively-sized animals initiated their tracking service by sniffing out lost hikers in the deep, icy snows of the 8000ft Mons Jovis Pass. A hospice for travellers was founded in 1049 by Saint Bernard of Menthon, and came to be named after him in the 16th century, along with the pass. The hospice later became famous for its use of St Bernard dogs in rescue operations.

The St Bernards were bred large enough to traverse deep snow and to scent out lost people. It is often said that they carried small casks of brandy around their necks (although this is only legend), in the belief that the liquor had medicinal properties.

The dogs work in pairs and use their fantastic noses in order to find the victim. Once a person is located, the dogs will dig away the snow until they reach them. Then the St Bernard will lick at the face and arms in order to rouse the individual. One of the dogs will immediately start barking in order to lead the rescuers to the location.

Thousands of people have been rescued by these wonderful animals. Supporters of the St Bernard claim that the animal can smell a human being from as far away as 1000 feet. They also say that if the wind is blowing in favour of the dog, it can pick up a scent from as far away as a mile.

We take advantage of these canine abilities every day, and each of us has a story about when our dog 'just knew' something. Take this example from Anna Osbourne, which gives a fascinating insight to her own dog's apparent 'sixth sense:'

"I was walking with my children and my mastiff, Presley, in a very large park close to our home. My son is a very small baby, and my daughter is ten, and we were playing a game of fetch with Presley in an open, flat field surrounded by trees.

"Something caught my eye from a distance. It was a man behaving very erratically and walking as if he was in pain or on drugs, jerking and unsteady. He was heading in the opposite direction, towards civilisation and out of the gates, so I thought no more of it and we carried on our game.

"It was becoming dark so we started to pack up. Presley was tired and happy, and so were the kids. The next thing I knew, this man was heading straight towards us. This was not normal human behaviour; there were many other routes around the park that didn't cross our path, and he was making no attempt to communicate with us, just walking, and I can only describe it thus, like a zombie intent on destruction and with no expression on his face.

"I didn't want to panic the children, so turned side-on to his approach and called my daughter to get ready to go home. Then my dog stopped her game and did something I've never seen her do before. She looked at this person heading toward us and, in less than a second, had assumed a herding position, moving my daughter back toward me. She then placed herself ten feet away from us facing the man, and began a perfect defence pattern, standing up on her hind legs, slamming back to the ground, and emitting a bark like nothing I've ever heard in all my years with her, all without moving an inch from her spot.

"The situation was now critical: the man was not responding or showing any sign that he had

CLEVER DOG!

registered what was happening. No one in their right mind would have kept moving with a Cane Corso in front of them displaying very clear signals that they were not welcome. Presley then began a circling pattern around him, still barking and slamming her feet on the ground, showing no fear at all, she was just very intent on getting her message across – she knew something was very wrong.

"I then shouted that the man should stop moving – he was still not acknowledging the presence of the dog, and heading toward us at speed. Presley was now inches from his face, still jumping but not touching him, and still barking, but not attacking. He finally stopped; he seemed to snap out of some state he was in – I can only assume he was on drugs. Presley stopped her antics but held him with a fixed stare, giving a low growl and looking to me for instruction.

"I called her back to me and she walked backwards to my side. It was this that made the man run, fast, the other way. It was the realisation, I think, that she was completely under control and still capable of defending us that made him see sense.

"The interesting thing about this story is, subsequently, we have seen the same man on the street, but 'sober,' and he has caused us no problem, nor shown any knowledge that he recognises us or the dog. Presley also ignores him, as on the street and non-aggressive, he is no threat.

"Presley demonstrated every great attribute the dog possesses that day: loyalty, intelligence, bravery, trust, and, subsequently, calmness."

Lessons we can learn from dogs

Dogs have an incredible ability to not only trust their intuition, making rapid decisions on the basis of gut instinct, but more often than not they get those decisions right. We, too, have these innate skills and abilities, but for too long have allowed them to lay dormant. Practice makes perfect, and learning to trust our natural instinct is one of the most valuable lessons we could ever learn from the dog.

Humans are very prone to over-thinking our decisions. There are too many external stimuli that can cloud our judgement. Gut instinct is there for a reason. We keep as memories the feelings we get when a combination of subtle stimuli are presented to us. Whenever we feel uneasy 'for no real reason,' or when we instinctively sense danger, this is our instinct at work, rapidly processing the combination of stimuli and running it past previous examples occasions when we felt the same way.

Dogs have this ability and still use it because their survival in the wild would depend on it. We humans are prone to ignoring our gut and re-analysing a situation that our brain and gut have already evaluated.

How does our dog know when someone is arriving at the house, even before the car pulls in? He's 'thin slicing' the information he has at hand. He can tell when visitors are expected because a frenzy of tidying goes on, alerting him to the possibility of people arriving. A dog will hear a car engine slowing as it approaches, which tells him that a car may be pulling into the drive. He gambles that this is the expected guest, and begins to prepare to greet the arrival before we've even realized anything is occuring!

Soldiers, police officers, and even professional gamblers use their instincts because their survival in the theatre of war, in the fight against crime, and at the poker table depend on it. We are capable of using our gut instinct when we need to – we just need to do this more ...

EIGHT

THE SECRET TO A HAPPY, HEALTHY, WELL-BEHAVED DOG

Gentle, effective dog care

"Properly trained, a man can be dog's best friend." – Corey Ford

We can't discuss the ever-evolving relationship between man and canine without taking the time to examine role reversal in this respect.

In this book we have looked at the dog from the point of view of what we can learn from him. Of course, it is our responsibility to guide and teach our four-legged friends how to show desirable, acceptable behaviour in our modern, human society. Sadly, in this endeavour, too many dogs are let down through naivety or simple plain ignorance.

Dogs come to us, in the main, hard-wired to learn and please us. Centuries of domestication, selective breeding, and what should be expanding knowledge of canine learning have resulted in an animal that is pliable, biddable, and extremely intelligent. In this chapter, we take a break from the notion of how we can learn from dogs, and focus instead on the crucially important aspects of how we can best serve them in terms of being their teacher and carer.

"He was the consummate protector. My most feared enemies dared not come around so long as this white-faced, fearless guardian was about. His persistence was unmatched, and his only satisfaction was to know that I was safe. A wise old boy, well travelled and experienced, he had instinctively come to sense the dangers long before I ever recognized them. His methods were fierce and swift; like a fire burning out of control, he consumed the prowlers at once, leaving no trace of the treacherous beings that crept up in the night.

"He joined my family in November of 1999, bringing smiles to our faces and joy to our hearts. Not unlike a firstborn child, he depended on me for all of his needs, and I would not come to appreciate how our roles would switch until much later in our lives. As we entered adulthood together, we experienced the trials of change. At first it was a new family member, our daughter, whom he took under his care from the moment he met her. A larger family now to protect and to care for, he carried out this role admirably for many years. In time, my wife and I – his 'parents,' as we affectionately called ourselves – decided to go our separate ways. I could see his heart breaking over this decision, one that he had no part in, one that would ultimately divulge the true purpose of his life's work.

"He was middle-aged when we moved to Florida. Although this stop on the journey through life was filled with morning and evening walks on the beach, one of his favourite places to play, he became very attentive to my need of him, staying closer to me and becoming acutely aware of my hidden enemies – feelings of regret, hopelessness and despair – the demons that would torment

CLEVER DOG!

me for the next several years. We left Florida and headed west, chasing that disappearing sun we watched together so many times on those grainy white beaches. Our brief time in the mountains unveiled a new bond between us, as if we had rediscovered our youth. We hiked, we swam, and we played in the snow but the cold nights would bring us both pain.

"He cared for me up until I found someone and something very special. A special person that he knew would not ever leave. She met with his approval, and he greeted her daily, just as he had done with me for so many years before, bearing gifts of all kinds. Usually a leaf or a stick, whatever was around, but many times something more precious to him, such as one of his many balls, or even a mislaid sock or similar article of clothing.

"Tucker fought hard with his tail wagging up until the end. He was a good dog, a great friend, and I will miss him tremendously." – Rudy Landry.

Rudy's is a familiar tale.

All throughout the world dogs and their owners share a bond that is unmatched between people and any other animal, and, in some cases, another person. Our dogs protect us, they make us feel safe. They entertain us, they comfort us, they enjoy us and we them. We witness their rapid learning curve and marvel at some of the incredible talents we often didn't realise they even possessed.

To truly get the most from our relationships with our beloved canine friends, we must understand them and appreciate what they need throughout their lifetime. From puppies – a gentle guiding hand to point them in the direction of what we'd like from them – to older dogs who may come to rely on us not just for specific care in their later years but, heartbreakingly, to sometimes make a soul-crushing decision about their continued existence.

Let's start at the beginning, shall we? Puppies!

"Who doesn't love a puppy? "There's no psychiatrist in the world like a puppy licking your face."
– Ben Williams

I think he was on to something, that Ben Williams.

Puppies tend to stir emotions inside us that we can't categorise, but just the sight of them evokes feelings of pure, unadulterated joy. Their fresh-faced view of a world they don't yet fully understand – yet which fills them with glee – has attracted us to them for centuries.

When we bring a young dog into our lives it's easy to get carried away with the novelty and pure exuberance of it all. I'll be brief here (this isn't intended to be an owner's manual, after all) but a few tips and insights to the basic, tried and trusted principles of kind, caring and responsible dog ownership can be a timely reminder of what our responsibilities are.

On a pup's first day at home, giving him a complete tour of the house on a loose lead can be a great 'start as we mean to go on' experience. This will be your pup's introduction to whatever limitations you want to impose on future access to your home and possessions (furniture, shoes, books and toys, etc). Now's the time for puppy to learn what he can and can't play with.

This is not the right time for 'no,' as he may come to think that this is his name! Instead, use a guttural "yack!" combined with a very slight tug-and-release of the lead as he sniffs about to warn him away from untouchables. He's new at this, but simply calling your puppy's name in a happy voice may be enough to get him to look at you, when you should then respond with a heartfelt "Good dog!"

Continue the happy chatter as you move on, talking kindly to him to reassure; remember, your puppy has only just left his dam and littermates to assume this dog-human relationship. Believe it or not, he's learned a lot already such as that if he plays too rough, his littermates (or dam) will often warn him of such by yelping. You can emulate this way of getting the message across if he nips too hard; instead of scolding him, which might frighten him, a yelp from you can be one of the most effective ways to teach him bite inhibition.

The human-canine teaching language is based on short, simple words that are consistently applied to specific actions. Dogs do not need a wide vocabulary of commands. It's very possible to teach a dog to the highest standard with a short repertoire consisting of just the following cues:

• No
• Come
• Heel
• Sit
• Down
• Stop

• No

The most over-used word in dog ownership, 'no' should be used rarely and it should be effective, conveying to a dog that he should stop whatever it is he is currently doing. If used correctly, the dog's attention will be diverted from whatever undesirable thing it is that he's doing, whereupon, we should immediately reward him for his compliance. If this doesn't do the trick, instead of repeating the command (which simply teaches him that 'no' doesn't actually mean 'no'), we should calmly and quietly approach the dog and physically distract him, either by drawing his attention with something more attractive to him, or by gently placing our hands to his side and focusing his attention on us, then rewarding him for it.

• Come

This is the recall command, and we'll briefly examine recall training techniques in a moment.

• Heel

Use this command to ask your dog to walk at your side, either on-lead or off.

• Sit

Self-explanatory but with a codicil. You may notice I have not included the cue 'stay' in the foregoing list. This is because I have always believed it to be an unnecessary command. If a dog is placed in the 'sit' or 'down' position, he should 'stay' in this position until given a further cue, rendering the 'stay' cue totally unnecessary.

• Stop

In many ways, this is similar to 'no' but with one specific difference; it should be used when a dog is in full flight. 'Stop' means exactly that: stop and don't move, which, potentially, can save the life of your dog in certain situations. If your dog is in full pursuit of a rabbit and is about to run head-first on to a busy road, are you confident that he'll stop if you tell him to? If not, you need to get working on the stop command. Note: the 'stop' and 'come' commands can be easily replaced with a whistle cue: one long toot for 'stop,' several short 'pips' for 'come.'

Familiarising your puppy with these cues early on in his life can pay dividends later on.

Socialisation

How important is socialisation? Well, according to leading British veterinary charity the PDSA, a staggering one in three of all dog bites is as a result of poor or improper socialisation. The charity recommends that owners properly socialise their puppy – to humans and other animals, and specifically other dogs – as early as possible to prevent them becoming a possible danger to themselves, other animals, and to people – incuding you.

Sean Wensley, Senior Veterinary Surgeon, says: "The 'socialisation period' is a critical time in a puppy's development, when the puppy learns how to interact with other dogs and people. This important time starts at about three weeks of age and ends at about 12 weeks. So, if you have just become a new puppy owner we strongly recommend that you see your vet for advice on socialisation as a priority."

He continues: "An increasing number of veterinary practices hold puppy parties to allow

CLEVER DOG!

your puppy to meet others in a safe environment while they're still young. The more your puppy experiences other dogs, people, sights and sounds, the more relaxed and well adjusted he will be as an adult."

If, by the age of six months, a puppy has not been socialised, it is unlikely that he will ever be, and, without this, fear of the unknown can lead to aggression in later life.

Another key factor in raising a well-adjusted and well-behaved dog is training. Puppies have the greatest potential for learning, but it's never too late to begin training a dog.

Tips for better canine socialisation
• A wide variety of friendly, vaccinated dogs, such as those belonging to family members or friends
• Young people and children (under supervision)
• Men and women of different ages
• People of different ethnic groups
• People using aids, such as wheelchairs and walking sticks, and carrying/using umbrellas
• People wearing different types of clothes, eg hats and scarves
• Household appliances, such as the vacuum cleaner and washing machine (introduce gradually)
• Travelling in the car – beginning with short journeys to nice places
• Being alone – gradually acclimatise to being left alone for increasing lengths of time, up to a couple of hours
• The noise of thunder and fireworks. Commercially available CDs can help a puppy gradually get used to sounds like these, as well as specially-made pheromone products
• Traffic and traffic noise, especially air brakes, horn, etc

Common canine behavioural problems
Without doubt, in my time as editor of *K9 Magazine,* and in my former life as a full-time, professional dog trainer, the most common behaviour problem I encountered would be inconsistent recall, typically, dogs who would be fine sometimes, but then at others would chase after the nearest thing that moved, with no regard for their owner's pleas to come back.

I could probably write an entire book on teaching an effective recall, but let's see what we can do here in terms of covering the most important elements of this often troubling, undesirable canine behavioural trait, using the most effective dog training tool I've ever encountered.

I began to train my first dog from the moment he came home with me. It was fun. It was, if I'm totally honest, pretty easy, too. He was a Labrador and his eagerness to learn and please me made training him fun, enjoyable, and, almost without exception, flawlessly simple. Then he grew up and things began to change. Luckily for me, by that time I'd discovered the most important dog training tool I've ever used. I swore by it then, I still do now ...

Jackson was my first dog. A handsome yellow Labrador.

As a puppy, I taught him to do lots of things: sit, come back, walk to heel, lie down, bark on command, give a paw – all so much fun, so easy to achieve. He was, in effect, and in keeping with the theme of this book, training me!

Then, almost overnight, he began acting like a teenager.

Probably because he was one!

More worryingly, I had progressed with his training to the point where I actually wanted to compete with him in working tests and trials. He had everything in his locker; he was fast, strong, intelligent – really intelligent – and he loved to work.

But his recall was, at best, 50/50.

If I'm totally honest, he'd only ever recall if the distractions and temptations around him were minimal.

Fortunately for me, I was learning the art of whistle training.

Fortunately for me, every dog I've ever trained since – regardless of breed, discipline, what level the dog was at – have all been trained using a whistle.

It's only now, at a point where I know more of the theory of canine learning that I appreciate just how and why whistle training is so incredibly potent.

The whistle, you see, is constant, consistent, emotionless, and incredibly easy to operate – you don't even need to charge it up, follow an instruction manual, or get a new one every other month.

They cost less than £10 each, and I still have the same whistle I acquired ten years ago, which I've used to train hundreds of dogs. That's real value, I'm sure you'll agree.

The principles of whistle training are mind-blowingly simple. You start at home, you give several short toots and always and without exception, you associate the whistle with great rewards. The earlier you start this in the dog's life, the better, but don't hesitate to adopt a whistle training recall strategy with a dog of any age. It works, it really does.

The main benefits of using a whistle:

• A whistle can be used by anyone! Now, I know that seems obvious, but think about it. Most family dogs hear many different voices throughout their daily lives, but a whistle sounds the same, no matter who is blowing it. The dog trained to recall to a whistle will do so regardless of who is blowing it, although there are ways in which you can make your whistle recall unique to you.

• A whistle lacks emotion. Ever tried to recall your dog when you're in a panic? Or a hurry? Or even when you're a bit angry? Think your dog can't tell? Think again! A whistle lacks emotion and it is consistent – something which is absolutely crucial to successful dog commands.

• The sound of a whistle carries a long way, not everyone's voice does. Besides, nobody wants to be the person at the park who's bellowing at their dog to come back. A whistle is a sharp, sophisticated way to communicate with a dog outdoors.

• Dogs love the whistle. If trained properly, the sound of a whistle can be as exciting to a dog as the sound of the biscuit tin being opened (yes, *that* exciting!). Believe me, my dogs go completely mad for the sound of the whistle and there is nothing – absolutely nothing – in the world that prevents them recalling when I blow it!

The importance of achieving a reliable recall can't be overstated. In fact, it's potentially life saving. I could probably take up a lot of your time discussing the intrinsic technicalities of recall training, but I just don't believe it should be that complicated. Using a whistle cuts through so much clutter. If you can get the dog to associate the sound of the whistle with a reward that truly motivates him, your chances of achieving a 100 per cent reliable recall will increase exponentially.

Other common dog behaviour problems
So, we've covered the top dog behaviour problem – recall – but what else tends to crop up regularly with the many thousands of dog lovers and owners I've had the pleasure of interacting with over the past couple of decades?

Pulling on the lead
Dogs like to explore scents, sounds and sights, and tend to pursue those interests with enthusiasm – even when that means vigorously towing their hapless owner down the street at the other end of the lead! When your dog is young and small, pulling might not bother you, but it may be a different matter as your dog matures, when it turns into a real problem, one that can deter many owners from taking their four-legged friend for a walk. This can result in the dog only ever being walked late at night when there's no one else around, and there's little opportunitiy for him to meet up with and interact with other dogs.

If your dog is large and powerful, you'd better fix this pulling habit early. Even if he's small, habitual pulling against the collar concentrates uncomfortable pressure on the dog's throat, causing gasping and wheezing, and can even collapse his airway and do permanent damage. It's not my intention to cause alarm, but the damage dogs can suffer as a result of constant pulling is genuinely serious and very unpleasant.

The good news is that teaching your dog to walk nicely on a loose lead isn't difficult – if you know a few tricks. Old-style training for loose-lead walking was based on jerking the dog's collar with varying degrees of force. But yanking a dog around by the neck can hurt it, and can also injure your shoulders, elbows, neck or back. It's not smart, it's outdated, and it does cause injury and suffering.

Fortunately, you can teach polite lead manners without resorting to this. A number of gentle, positive techniques for teaching loose-lead walking have been proven to work when consistently applied.

Starting out right with loose lead training

Believe it or not, most dogs pull on the lead because their owners inadvertently train them to. Yep, really; in a misguided effort to control their dogs, many people keep the lead short and tight, which teaches the dog to pull by holding him at a point where there is constant tension on his collar.

The taut lead makes tightness the norm for when you and your dog walk together. With a strong breed, this very quickly develops into an intense and physically demanding pulling on the lead.

Several types of leads can be used to change to a loose lead heel walk:
• 6 foot leash: this can be used either shortened or full-length, and is long enough to tie to your belt for hands-free walking
• 4 foot leash: similar to the 6 foot leash but less versatile
• 10 to 30 foot long line: your dog can learn to walk without pulling on any length lead. The long line allows safe control whilst giving the dog freedom to explore
• Retractable lead: these are handy, but, as they're operated by the dog pulling, they reinforce (reward) pulling on the lead, which counters what you're trying to teach. Be very wary of using this kind of lead until you have truly mastered the loose lead heel walk.

The absolute simplest method of preventing a dog from pulling is to constantly vary your pace of walking, in some cases, almost to the point where you take one step every 5 seconds. Stop regularly and ask your dog to sit. Change direction rapidly, even when out for a walk in public; yes, folk might think you're a little odd, forgetful even, when they observe you walking forward and then suddenly turning round and walking in the opposite direction, but you won't get arrested (probably)! The constant change of direction will alert your dog to the fact that you aren't necessarily headed where he thought you were ... Be upbeat in your walking style and reward constantly for walking loosely. The very second you feel any tension on the lead, stop or change direction immediately.

Separation anxiety

Along with pulling on a lead and not coming back when called, the most common canine behavioural complaints tend to be connected with separation anxiety, which can manifest as destructive behaviour in the home, barking/howling when left alone, and even self-mutilation.

Tim Miles, former Chief Veterinary Adviser to Britain's largest animal charity, the RSPCA, says: "It has been estimated that more than one million dogs become distressed and display anxiety when left alone. Typically, they will bark, chew furniture, and/or urinate and defecate indoors when left alone by their owners.

"Recent studies have shown that as many as half of all pet dogs may show signs of separation-related behaviour at some time during their lives. Dogs of any breed, size or age can exhibit separation anxiety, a behavioural problem which causes them to suffer.

"Separation anxiety is a common reason for dogs to be handed over to rescue organisations, but dogs obtained from rescue centres are no more likely to develop this behaviour than dogs obtained from breeders."

It may not be obvious to owners that their dog suffers from separation anxiety unless they find evidence of destructive behaviour, or toileting problems when they return home. If a dog has this problem it's important to let them know it's okay to be left alone.

Top tips to deal with separation anxiety:
• Be calm, consistent and predictable with your dog
• Reward good behaviour with your attention or with treats

• Ensure you start and end all interactions with your dog; don't allow him to
• Ignore any attention-seeking
• Never punish your dog as this will harm your relationship and may make your dog scared of you
• If your dog looks 'guilty' he has simply learned that you are sometimes angry when you return home. Your dog is not able to associate your anger with something he did while you were out
• Leave your dog alone for gradually increasing periods of time. To begin with, just a minute or so, gradually increasing the time you are away
• Give your dog something nice to do whilst you are away. For example, leave a tasty chew or a durable rubber toy filled with food

Research commissioned by the RSPCA's veterinary department in 2003 concluded that a behaviour test can be used by animal rescue centres to successfully identify dogs which are likely to suffer problems when left on their own in their new homes.

The test involved:
• Getting to know the dog and getting the dog used to the test room
• Assessing the dog's behaviour in the kennel
• Assessing the dog's behaviour when left alone in the test room for a short period
• Contacting every new owner 12 weeks after a dog was re-homed to find out how the dog was behaving

In 86 per cent of cases the test was effective at predicting which dogs would not react badly to being left alone. It was effective at predicting which dogs would react badly to being left alone in 72 per cent of cases.

As separation anxiety is such a leading cause of dogs being given up to rescue centres, it's crucial for us as caring, considerate dog owners to understand the causes, symptoms and treatment for the condition as best we can. If you happen to know of someone whose dog might be exhibiting signs of separation anxiety, perhaps it might be a nice gesture to offer some insight from the tips contained here, as such simple, easy-to-follow steps could potentially prevent another poor dog ending up in the already overcrowded rescue system.

Special bonus: a secret, little-used technique that can dramatically improve the relationship between dog and owner

Want to know a secret?

There is an incredible technique that dog owners can use to help them forge a truly special relationship with their canine companion.

It's almost impossible to over-emphasise just how criminally under-used, yet fantastically effective canine massage can be in terms of strengthening the bond between dog and owner, as well as providing a hugely beneficial health boost to dogs of all ages. So here's a brief introduction to the art of canine massage. You will be truly amazed, once you've mastered the technique and your dog has enjoyed the results, at how this simple, hugely enjoyable exercise can dramatically improve the bond between dog and owner.

An introduction to canine massage

This type of 'bodywork' can give your dog instant comfort and relaxation, and the feeling of being loved. When carried out on a regular basis, massage helps minimise sensitivity to touch (by you or a stranger), which makes grooming and handling easier, especially should he need to visit and be treated by the vet.

The growing popularity of animal massage (and other forms of complementary therapies) means that therapists are widely available in most countries, and especially in major cities. However, you can learn the basic techniques yourself and perform them on your dog, making the experience more enjoyable for you both.

Treating your dog to a relaxing massage is an excellent way to bond with him: not only will he love the way it feels, but will also appreciate the time and attention you are giving him.

CLEVER DOG!

There follows a simple, step-by-step procedure for a relaxing massage, starting from the top of your dog's head down to his muzzle.

Canine massage step-by-step: head to muzzle
• Call your dog to you and ask him to sit
• Ease your way gradually into the massage by caressing him lightly as you would normally do. That way, he won't think you are about to do something different that might alarm him and make him wary
• Gently caress him all over from head to toe, using long, firm strokes. This will prepare his muscles for more intense work, and has a relaxing effect. Do this for about two more minutes
• Put your hands over his head, then move one hand gently underneath his muzzle whilst your other hand strokes the top of his head
• Next, using circular movements, massage his ears, one at a time. The ears contain a lot of nerve endings, and massaging them will feel very good to him. Do this for at least a minute each ear
• Lightly massage his forehead for about 30 seconds, sliding down to his cheeks also for 30 seconds, and then to his muzzle for another 30 seconds

Canine massage step-by-step: neck to toes
• Begin with a neck rub. This will release a lot of tension in that area. Slide your hand down to the back of his neck (it's easier to locate the muscles here as the skin around that area is much looser). Using your fingers, gently rub the muscles up the back and sides of his neck, but ensure you're not putting pressure on the windpipe
• Consult your vet before attempting the following massage to ensure that your dog doesn't have any underlying back problems. With both hands, gently massage the areas each side of the spine. You must only apply the lightest of pressure, only slightly firmer than you would when stroking your dog. Do this for about a minute, then rub your palms along his sides for another minute
• Next, lightly 'walk' your fingers along both sides of his spine
• Help him lie down on his side. With gentle pressure, slide your hands down the top portion of his legs about five times. Then, while supporting each of the upper legs in turn, cup your hand around the thigh muscle and hold it. You needn't do anything more than hold the leg, as the warmth of your hand will warm the leg
• Lift each paw one at a time, and gently rub on and around the nails and paw pads in small, circular motions
• Carefully flex and straighten each of your dog's leg joints several times, working upward from the paw to the top of the thigh. Be careful to provide adequate support to the rest of the limb when doing this. Only move the leg through its natural range of movement
• Now, turn your dog onto his other side and repeat the previous three steps with the other two legs
• The final step is to apply long, gentle strokes, from the top of his head to the tip of his toes. Do this several times

Your dog has probably fallen asleep by now. If so, then you did a great job. Even if he's not sleeping, he's sure to be very calm and relaxed.

Given how little-known and under-used this superb technique for achieving a happy, healthy, well-bonded dog is, I strongly encourage further investigation of canine massage. You will learn all you need to know by picking up a copy of *The Complete Dog Massage Manual – Gentle Dog Care*, by Julia Robertson, a qualified therapist (published by Hubble & Hattie).

Lessons we can learn from dogs
As already mentioned, this chapter has been the exact opposite – in terms of general theme – of the rest of the book. But, when you think about it, we DO learn from dogs the more we attempt to understand them, and the more we work with them in terms of shaping their behaviour to suit our lifestyles and our modern society.

THE SECRET TO A HAPPY, HEALTHY, WELL-BEHAVED DOG

Dogs come hard-wired to want to please us, yet many dog owners struggle with teaching desirable behaviour to such an extent that rescue centres the world over are full to the point of bursting with animals that have been given up on.

If we are custodians of the dog, we must surely owe it to him to properly understand what makes him tick and what we can do to promote health, happiness, and behaviour that lets him live within our society, but which also doesn't impinge too much on his natural proclivities. If there is a life lesson to be learned here, it's this: the dog does not come with an instruction manual, but will happily exist side by side with us if he is given well defined boundaries, positive rewards for his good/desirable behaviour, and is exposed to appropriate levels of mental and physical stimulation to match his particular breed type and lifestyle requirements.

Owning a dog should be a privilege, not a right. We should remind ourselves how lucky we are to have a creature by our side that would lay down their life for us if needed. We have a friend who doesn't care whether we are in debt, whether we are struggling with work, or whether we wear the latest fashion in clothes.

We strive for unrealistic things from our relationships, yet sometimes overlook the loyalty and true friendship that dogs offer us. Then, should we find that dog ownership becomes somewhat inconvenient, there are those, sadly, who simply hand the dog to the nearest rescue centre, in the process losing all that a dog gives, every day of his life.

If people received the same treatment, no doubt we'd be jaded and bitter. If a dog is lucky enough to find a new home, however, he isn't suspicious or cynical; expecting that the same bad pattern of events. No, our canine friend grabs the opportunity for happiness with all four paws and a furiously wagging tail, determined to make the absolute best of his new life!

COMMUNICATION: DOG STYLE

The art of talking without speaking

"One reason a dog can be such a comfort when you're feeling blue is that he doesn't try to find out why." – Unknown

If modern man met a dog 10,000 years ago, because a dog's capacity for communication was (and is) untroubled by etiquette and language, the dog would be able to communicate effectively with that person more successfully than people of totally different backgrounds could communicate with each other.

Dogs are highly sociable pack animals, with acute senses that enable them to smell, see and hear things much better than can humans. Their ancestors used communication to get along, hunt successfully and raise their young; survival in the pack depends on the ability to communicate with each other, and dogs derive a great deal of information from the messages of other dogs.

There are no language barriers in canine communication; all dogs speak the same lingo. They communicate with each other and with humans in many ways, which includes body language, facial expressions, scent and vocalisation (barking, whining, yapping, and growling). Unlike humans, dogs don't need to explain something, they simply show what they are feeling and what they want.

Many dog owners are familiar with the sounds their dogs make, and have an idea what it is their dog is trying to communicate. However, a dog's movements and actions – his body language – still baffle many dog owners.

Like humans, dogs can communicate and signal to each other with almost every part of their body. Generally, canine communication is 'voiced' via the eyes, ears, mouth, tail and body position and posture. It's worth noting that not all dog breeds are wolf-like in conformation, and we have yet to fully appreciate whether breeds with droopy ears, docked tails or long hair (unable to demonstrably raise their hackles) are at a disadvantage when it comes to signalling behaviour.

Through their body language dogs can convey many emotions – anger, excitement, fear, playfulness, dominance and submission, for example. They can provoke a fight or avoid one; dogs are experts at reading the most subtle nuance. Interpreting individuals' body language is a complex affair, and there are no hard and fast rules to understanding what dogs are trying to say to us humans. A wagging tail may indicate a happy and relaxed animal, but with another can signify anxiety and stress. In fact, tail wagging is one of the behaviours that we humans misunderstand the most. For example, we often mistake a raised and wagging tail as a sign of excitement, when it can also indicate fear, stress or irritation, which is why it's important to take in the dog's attitude as a whole (just as other dogs would do) when interpreting behaviour.

Different states and attitudes show in a variety of ways; here's an idea of some you may witness:

Happy
A happy, carefree dog will have a relaxed body and free-flowing movement, with his weight equally balanced on all four paws. His head will be held high, his tail will be held down (but not tucked under or between his legs), or wagging side-to-side or in a circular motion. A happy dog's ears will be relaxed and the hair on his back will be lying down. Eye contact is minimal, and his nose will be working. A happy dog will have his mouth closed or slightly open.

Excited and curious
As with a happy dog, the body will be relaxed with lots of spontaneous movement; the tail will be upright and wagging swiftly. The dog's ears will be perked and facing forward, and his eyes will be wide open. The dog's mouth will be open, and some seem even to laugh or grin when the corners of his mouth are turned slightly upward.

Submissive
A submissive dog is not a fearful dog. Submission signals are sometimes referred to as 'distance reducing signals' in that they encourage the other dog(s) to approach, thus minimising any threat and ensuring harmony in the social interaction. Submission can be divided into three areas: passive submission, active submission, and play.

Passive submission
In a passive submissive state the dog's body is low to the ground. In increased submission the dog will roll over on his back, exposing his belly, with head low and possibly turned sideways. Eye contact is avoided. The dog's tail is also low and may be wagging. Ears will be down and flat against the head.

He may raise his paw to solicit play behaviour, or in defence to ward off the other dog. Submission is also displayed by licking the other dog's chin. Extremely passive submissive dogs may urinate, especially puppies.

Actively submissive
This is more common behaviour than passive submission. Head and tail are held high on the approach, and, once the other dog has been reached, the first canine assumes a more passive submission state; lowering his head and body, and diverting his gaze. Licking and nudging with his mouth is also common behaviour.

Play behaviour
Play amongst dogs can include elements of fighting, sexual, and predatory behaviours. The play bow – when the dog lowers the front of his body, whilst holding his rear end in the air – is one of the most common body postures that dogs use to solicit play. The play bow is only used in play behaviour.

The dog's body movements are bouncy and jerky. The raising of a paw can also signify play; other indications include erect ears, a wagging tail, and greeting 'grin.'

Dominant
A dominant dog is not an aggressive dog. This fella has a confident and assertive attitude to the situation at hand. In a dominant state 'distance increasing signals' are exhibited to communicate a 'go away' message. The dog makes his body appear as large as possible, his posture is high and tall, hackles may be up, his tail is erect or straight out from the body; it may even be fluffed up (tail height gives a good indication of confidence level). The dog's ears are standing upright and facing forward, and his eyes are wide open and staring; mouth is usually closed.

Dominant dogs tend to rest their heads across the back of more submissive pack members.

CLEVER DOG!

Aggressive

As with dominance, aggression also involves 'distance increasing signals;' therefore, the dog's body is tense and upright, and his hackles are raised on the neck and back. Usually, the tail is held straight out from the body or high, and could be fluffed up. Eye contact is rife and focused on the target – therefore, a key indicator of threat behaviour. In an aggressive state the dog's ears are usually laying flat back on his head, although could also be erect. As aggression intensifies, his lips are drawn back, exposing the teeth.

Alert/interested

This is a common state when dogs meet for the first time, and continues for only a short period as each decides whether to exhibit dominance, submission, or play behaviour. Posture is slightly dominant – upright, with the weight spread evenly on all four paws. His eyes are wide open and focused, and his mouth is usually closed. The tail is up high and possibly wagging, and his ears are up and cocked towards the target.

Fear/threatened

A fearful dog instinctively protects his belly by crouching low to the ground. He makes himself appear as small as possible. Shivering and trembling is also a key indicator of feeling threatened, and he may be 'frozen' in place. His tail is usually between his legs or held low, and his ears are laid flat on his head. Eye contact is avoided and he may show a half-moon eye (the whites of his eyes are visible). Other signs include yawning, licking his nose, or biting himself.

Signs which can indicate stress

A dog can exhibit many physical signs which indicate stress behaviour, and these occur most during moments of emotional conflict and during a fight or flight response. Dogs use these behaviours as a way of calming themselves, and they are sometimes referred to as 'calming signals.' The signs themselves may seem inappropriate to the situation – yawning when not tired, or panting when not hot. Other stress signals include licking lips and nose, body scratching, shaking (as if wet), looking away, and avoiding eye contact, and possibly even stretching.

Body language is not the only way that dogs communicate with each other; vocalisation and olfactory signals are used extensively in canine conversation.

Vocalisation

Vocalization is used alongside body language to communicate. Dogs can emit different types of sounds, including barking, groaning, growling, grunting, hissing, howling, panting, puffing, screaming, whining, whimpering, and yelping. All of these sounds are used in different situations.

Using sounds starts from day one; puppies make a mewing sound, calling for their mother in search of food and warmth. As they get older the sounds become more varied and louder.

Barking is the most apparent of all the vocalisations, and usually begins when a puppy is three weeks old. Barking does not always signify anger, and the tone of the bark can reveal a great deal. High tones indicate play behaviour and excitement, for example, whilst lower, deeper tones indicate distancing behaviours such as defending territory or alerting others to a threat. Dogs also bark when they are bored or seeking attention.

Whimpering and whining is used by both puppies and adult dogs, and is a way of communicating frustration, pain, distress, submission, or – very commonly – attention seeking.

Howling is usually associated with pack situations, and is the canine method of long-range communication. Dog packs and wolves use howling to reunite the pack.

Growling begins when the puppy is around a month old, and is sometimes used to warn another dog not to approach; also to show dominance. However, it is not always associated with aggression and threat as it can be used during play, too.

Grunts indicate contentment, sometimes heard during stroking and moments of affection.

As with body language, sounds can have different meanings, depending on the situation, which is why it's so important to study both vocal and physical signals when interpreting behaviour.

Olfactory signals

Dogs have over 200 million scent receptors, which provide an extremely powerful sense of smell (humans have just 5 million). Generally, the bigger the dog and the longer its muzzle, the more olfactory receptors it has.

Dogs can detect smells that humans don't even know exist, and can establish a lot about another dog through scent: what sex it is; whether or not he or she is neutered; friend or foe, and even the age of the other dog. Through scent, a dog can also pick up on the other's mood and social status.

Dogs leave their own scent for other canines, either by urinating, defecating or rubbing themselves on objects or each other. This is known as scent marking and is commonly used to attract sexual partners, mark and maintain territory boundaries, or establish a social ranking. Urine marking is more common in male dogs; female dogs tend to use urine marking to attract a mate. The scent in a dog's urine is as unique as human fingerprints.

The scent of a human is also very powerful; a dog will remember some people with affection and fun, and others with fear and intimidation. When humans cry, our tears contain different hormones depending on whether they are 'sad tears' or 'happy tears.' Amazingly, experts believe that dogs can smell these hormones and tell the difference between the two.

Canine understanding of words

Have you ever wondered whether your dog understands what you are saying to him? Well, dogs do have a profound mental capacity which allows them to associate certain sounds (words) with particular actions. Pavlov's experiment involved ringing a bell just before giving his dogs some food; after several repetitions, the dogs would salivate upon hearing the bell ring, irrespective of whether they saw or smelt food. Through this form of repetition and conditioning, the dogs learnt to associate the bell with food, and this is also how they recognise our words, by associating a word (sound) with an action.

It's been said that the language comprehension of a dog can rival that of a two-year-old toddler. The psychologist Stanley Coren reckons the average 'trained' dog knows about 160 words, including signs, signals and gestures. Coren has also said that the average dog can count up to five.

German researchers have discovered a Border Collie – Rico – who knows over 200 words. Rico has the astounding ability to select unfamiliar objects with a name he has not heard before from a group of known objects.

Although dogs can't understand our language in the way that we do, certain studies have shown that dogs understand human contextual cues and gestures more so than other animals, which means that our body language is a big factor when communicating with dogs, and the reason why dogs don't respond as well when listening to sounds through an intercom or via a telephone.

Have you stopped to think of all the body actions you use when giving a cue to a dog; for example, 'go outside' normally involves pointing to a door, whilst 'off' and 'down' can also involve pointing. To dogs, a word is just a sound, although they are able to distinguish between words that are very similar apart from a different vowel; for example, 'dog' and 'dig.' However, dogs are unable to spot a change in a consonant, so saying 'fly down' is likely to result in the dog laying down, for example.

Dogs respond to their name not because they appreciate that this is their name, but because they understand they have to react when hearing this sound (their name).

Deaf dogs

All dogs are born deaf, and so their sense of smell is the first used. Next, they begin to see, and

CLEVER DOG!

hearing comes last. For humans, the reverse is true; we first hear, then see, and then smell. As dogs are born deaf, a deaf dog doesn't know that he is deaf, or that the rest of his littermates will soon be able to hear. His world is what it is and he has the ability to live just as happy a life as a dog who can hear; therefore, deaf dogs get on much better than we might think, not worrying about the past or the future, and living in the moment, unlike humans, who have the ability to dwell on their disabilities.

Deaf dogs rely solely on their senses of smell and sight, both of which become even more sensitive in the absence of hearing. Another sense which humans lack is the ability to read energy. When a human is scared or anxious a dog will be able to discern this; likewise, when a human is confident and in control, a dog has the profound ability to pick up on this energy. In an ideal world we would communicate with dogs through body language and energy: verbally, less is more.

Many behaviour problems associated with deaf dogs are due to a lack of leadership in the owner, or weak energy that the person is exhibiting, notably 'pity' for the dog and his disability. Therefore, we should never feel sorry for a deaf or disabled dog as they don't feel sorry for themselves, and this weakness will be considered an opportunity for them to take over as pack leader.

It's natural to assume that just because a dog is deaf, he will be difficult to train, or even exercise off the lead. This assumption is wrong; with a little hard work and a lot of love, imagination, patience and leadership, you can train a deaf dog just as well as his hearing buddies.

Hand signals are a great substitute for words when training a deaf dog. You can use actual sign language, or make up your own to suit you and your dog – for example, clapping your hands when he has done something good. A great place to start is with the signs and signals you would normally use if you were speaking to a hearing dog. The signals need to be clear and distinguishable from each other and, most importantly, consistent. Positive reinforcement should be used to support the training.

The first sign you should teach your dog is 'good dog.' An easy way to do this could be clapping your hands, another signal is a thumbs up (thumbs down could signify 'bad dog'). You will then want to work on signals for the following commands: sit, down, come, stop, and no. Some dogs have been known to recognise up to 50 signs.

The key is to train the dog in the same way that you would train any dog – through repetition and consistency. With lots of positive reinforcement, a deaf dog can learn signs and signals in exactly the same way that a hearing dog can learn words. As dogs are experts in reading body language (including facial expressions), it's recommended that we still talk to deaf dogs whilst giving signs. Even though they won't be able to hear us, they will learn to read our facial expressions, which you should exaggerate, smiling when you are pleased or rewarding, and scowling when you are not happy. In this way, your signs will be learnt a lot quicker and will also have more meaning.

Vibrations can also be picked up by deaf dogs. If your dog is not looking at you and is across the room, try stamping your foot on the floor. Vibrating collars are also used to get a dog's attention.

Recall is a much harder command to train in deaf dogs, and one which requires much more patience. The most effective and safest way is by using a vibrating collar, as you can train your dog to come back to you when he feels the vibration. These collars only work up to a certain distance, so only let your deaf dog off the lead when his recall is perfect; training with a long line in tandem with a vibrating collar is a great way to practise his recall. Using a flashlight when outside in the dark is another way to get your dog's attention. Sneaking up on a deaf dog when he is asleep should be avoided; make sure you always wake him gently. A good tip is to place your hand in front of his nose so that he can smell you.

No one should give up on a deaf dog; he can easily live a happy life through great leadership, consistency and love. Another important tip is to make sure his ID tag states that he is deaf. See *Further reading* for more information.

Blind dogs
Blindness in dogs is a slightly more challenging disability. Dogs that are born blind have the distinct

advantage of not being aware of their disability, and so will cope quite easily. Hearing, smell and touch will get them through, provided their owners are able to cope with not being able to communicate with them visually.

How you are with a blind dog need not be that different, really, when you consider that giving affection relies on touch, giving cues relies on sound, and feeding relies on smell, and, to a lesser extent, vision.

There are possibly issues for when your dog is left to his own devices, though even he will have developed many coping mechanisms without realising it. For example, a blind dog will know his way around the house based on smell and hearing, and, if loss of sight occurred during his life, his memory also. It is only when we present him with unnecessary obstacles that problems arise. Moving the furniture around might not cause problems for a deaf dog, but for a blind dog this could take a lot of getting used to, as regular, known paths and routes are no longer there ...

Dogs with little or no vision also need slightly more supervision to help them cope with the ever evolving human world. In the wild, a blind dog would inevitably be removed from the gene pool by a predator. The blind dog's ability to cope with predation, while not totally non-existent, would always be rather less than the rest of the pack. For a domesticated animal, this is not an issue, but we must be mindful of dangers that can't be detected audibly or by smell. You can train a blind dog to stop and sit whenever he gets near to a road, but no right-minded dog owner would give any dog free-reign when close to such a danger. See *Further reading* for more information.

Body language

Some people claim to be better at reading dog body language than that of people. Sometimes, we don't even understand a fellow human's words, even though he or she is speaking the same language!

The human language is open to misinterpretation as we communicate on many levels and with many people. We may be blunt and direct with some, or ambiguous and fluffy with others. Our emotions very easily come into play when we communicate, affecting what we say and how we say it. Sometimes, we unintentionally offend others by our tone of voice or choice of words; other times we can cause laughter with our wit and humour.

Sometimes, small talk is what's required, but this is something that dogs don't do.

It's not that dogs are not emotional, they just communicate in a much more straightforward way. Dog communication is more honest and less likely to lead to confusion with other dogs, whilst it may seem ambiguous to us, between each other, all forms of canine communication – verbal and non verbal – are easily read and understood.

Dogs give clear signals to signify what they feel and expect of each other; they can be both direct and very polite, almost extremes, but always clear.

We can teach our dogs a variety of different things, but did you realise that dogs can teach us a thing or two? Things that could help us live happier, healthier and more fulfilled lives.

Dogs live in the here and now, they don't dwell on the past or worry about the future. Every day is broken into many worry-free, enjoyable moments. People, on the other hand, have a tendency to live in the past and worry too much about the future, while life just passes them by ...

Dogs use their instinct and go with their gut feeling. Humans have survived through centuries by relying on their intuition, but nowadays we think and analyse too much and rely on what others tell us. We reason every decision, something that dogs are not able to do. It pays to trust your instinct sometimes.

Dogs are team players, get great joy from being with others – whether human or canine – and are eager to please everyone. They take direction well and are natural followers; many people, on the other hand, dislike being told what to do.

Dogs are not deceitful and are unable to lie to us or each other. They don't pretend to be something they're not, with them what you see is what you get and we love them for who they are.

Dogs show no prejudice; they don't care about our race or religion, how rich we are, how

CLEVER DOG!

good-looking or unattractive we may be, or if we have a disability. Amongst their fellow dogs they aren't bothered that Lucky is a mongrel, or Sammy a big German Shepherd, and they certainly don't judge Rocky because he looks like a 'pit bull type.'

Dogs don't hold grudges; if there's conflict in their pack they deal with it straight away and move on. People, on the other hand, have the tendency to hold grudges, feel resentful, and want to 'get even.'

Evil and jealousy don't exist for a dog, and the world would be a much better place if they didn't exist for us, either.

Lessons we can learn from dogs

Dogs are not called 'man's best friend' for nothing. Our canine friends stick by us, even when our human friends may not. They offer loyalty and unconditional love; when relationships fail, families move on, and friendships change, our dog is there by our side through it all. A dog shows us just what loyalty, faithfulness and kindness really are.

Dogs have basic needs. The dog who belongs to a homeless person isn't worried that he has to sleep on the street, he's happy because he's always got company, new people to meet, and is not cooped up in a house.

One of the main problems we have as dog owners is that we judge dogs by our standards. We want to give them what makes us happy; that's why people get rich selling pet jewellery! If we were to try and find out exactly what makes our dog happy, we'd realise that they don't want gifts or expensive food, but simply our love, attention and respect.

DOGS AND THE LAWS OF ATTRACTION

What dogs teach us about winning
friends and influencing people

*"We long for an affection altogether ignorant of our faults. Heaven has accorded this to us in the
uncritical canine attachment."* – George Eliot

Dogs have an amazing can-do attitude when it comes to adaptation, and their superior skills in
scenting, learning, and fitness have been used – and more than occasionally exploited – by us
humans since the very start of our relationship.

Dogs have proved themselves more adept at recognising small human gestures such as
pointing, glancing, and bowing, than even our closest relative, the chimpanzee. Studies at the
Hungarian Academy of Sciences showed that pet dogs consistently performed better at interpreting
these signals, which could account to some extent for their amazing success at ingratiating
themselves and satisfying our human need to communicate – making the dog a near perfect
'friend.'

Their fitness levels make them ideal working animals for rough or arduous terrain, and even
the owner of a companion animal can appreciate that having a dog can be akin to having a
personal trainer 24/7 – "Are we going out? What are we doing? Can we do running? I like running!
Faster, faster! Let's go up that hill/into that pond/all the way over there! Woo hoo! I found something!
Wanna see? I found a smell! Follow me, follow me!"– and all whilst displaying that most wonderful of
doggie traits: sheer joy at being alive and out in the fresh air discovering new things.

As a species we would do well to emulate this enthusiasm, finding joy in whatever we do and
striving to learn more. There's satisfaction for owner and dog in learning a new skill that they can
enjoy together; I learned to ride a bicycle as an adult just so I could facilitate my dog's desire to
patrol large areas of land, and it's benefited us both. We're fitter and I'm less grumpy, and we get to
experience the outside world at a pace which suits her and that I can keep up with. You might also
use rollerblades, skates, go-karts, or running shoes as ways to match the pace set by your four-legged
friend.

Dogs in the arts
The earliest cave painting showing some sort of domestic canine is at the Bhimbetka rock shelters in
India, and could well be one of mankind's first attempts to document the aesthetic appeal of the
dog (or it could be the first instance of advertising dog leads ...).

Cerberus, the infamous guardian of the gates of Hades, is the canine that immediately springs
to mind when we go back through visual history to a mythological age, where simply getting out of

CLEVER DOG!

bed in the morning meant taking your life in your hands. Often incorrectly depicted as having three heads (this was, in fact, the party trick of his sister, the Hydra), Cerberus did have a few of his own ...

From *The Theogony of Hesiod* (translated by Hugh G Evelyn-White, 1914): "... and then again she bore a second, a monster not to be overcome and that may not be described, Cerberus, who eats raw flesh, the brazen-voiced hound of Hades, fifty-headed, relentless and strong."

The appeal of dogs like Cerberus is that of the tireless protector, the fearsome yet loyal hound who will, without question, serve his master to the death. He is also partly responsible for the mythical Roman War dogs (even though Cerberus was Greek, the Romans were all about embracing other cultures) so often cited by Mastiff breeders as heralder of their bloodline, mentioned by Julius Caesar, no less, when he 'discovered' them on his jaunt to Britain. Legend has it that he was so impressed, he took some home and created the War Dog, an armoured, front-line animal that struck terror into the hearts of the enemy. Sadly, this morsel of 'common knowledge' is simply not true, and websites bearing an official 'translation' are actually quoting a misinterpretation from an 18th century novel.

Julius Caesar never mentioned Mastiffs in any of his writings, and neither did his generals. They mentioned Deer Hounds, and terriers, but not Mastiffs. Nor are there any records (and the Romans were pretty good at record-keeping) of any Mastiff returning with Ceasar from Britain. No canine armour has ever been discovered, although the canine flock guardians who ran with the legions would have worn spiked collars to protect them from wolves, which is probably how the myth originated.

No Molosser-sized-skeletons have ever been found in or near war graves (a Deer Hound was buried with a Roman archer in Chichester, but I'm not sure he counts as a War Dog), and there's no mention by the Romans or their enemies of big dogs ever being used in warfare. They mention their toilet block arrangements and how many prostitutes accompanied the army, but not any dogs, except those taken to guard the legion flocks or assist with hunts (nearly all Deer Hounds) which would suggest as strongly as it can be that the War Dog did not exist.

The reason I feel sure of this is that a colleague and I have spent many years, and countless hours of discussion with experts on Roman culture, Latin translation, and dogs, trying to prove that the Romans did indeed import the Mastiff from the UK, as opposed to the more likely story of it being a descendant of the massive hounds of Assyria (which doesn't sound as good as 'Molosser,' hence the second error in common knowledge, but then, that's another book entirely ...)

From the serene, pastoral scenes of dogs going about their business, tiny lapdogs secreted about the skirts and sleeves of Tudor ladies, beloved pets captured in portrait with fidgety children, right up to startling and unnerving hunt scenes such as *Hounds attacking a Stilt bird* by Franz Snyders, dogs have appeared in art since the day we became acquainted.

I was lucky enough whilst researching this work to discover the director Kath Burlinson, who has recently finished a run of a work entitled WOLF at the 2010 Edinburgh Festival.

This is a show set in caves, with the audience standing for 60 minutes whilst performers who had been extensively connecting with their inner-wolf, and examining the lessons and lore that these creatures have given us, move around, between and frequently within the audience, displaying sometimes uncomfortably close and intense body language and eye contact. The soundtrack is provided, in part, by text from Iain Finlay Macleod and composer Kerry Andrew, and partly by the cast. The performers both embraced and released an energy that we have been sharing with dogs and wolves (our own societal structure being more similar than many would openly admit) since we first learned to draw inspiration from them, and weave their tales into our own. Yet, because our busy, hectic lives mean that many of us spend more time on practicalities rather than creative or imaginative pursuits, we have been consciously suppressing elements of that energy which are actually important to us, not just in terms of our relationship with dogs, but also our relationship with ourselves and the world we live in.

Kath explains where her inspiration came from and how the work developed:

"Someone in a workshop once told me when they looked into my eyes they saw a she-wolf.

"That chimed with the fact that I had been re-reading *Women who run with the wolves.*

I decided to begin an exploration of wolf/human relations in myth, fairy tale, psychology and ecology. I gathered a group of actors and we did some exploratory weekends when people were available. We explored our own animality by examining our senses – especially smell and hearing rather than sight, on which we rely a lot.

"The project evolved over 18 months until we took it to Edinburgh Fringe. In the course of the development we visited the Wolf Conservation Trust near Reading to see live wolves. We investigated wolf spirits in Native American lore, and did research on wolf reintroduction and its effect on the environment. I consulted an eminent professor of biology at Imperial College; we read eclectically: novels and works of philosophy as well as biological studies. We were interested in how wolf packs organise themselves, their codes of behaviour.

"This was interesting to us, an ensemble, the intuition of wolves, their capacity to work together, to have a sense of the pack without foregoing their individuality ... we humans often seem to find that balance hard.

"The cast was put through intense experiences, confronting sometimes alarming aspects of their own sense of self and belonging, as well as a group of real wolves. Whilst it may seem extreme to suggest that reconnecting to the wild aspects of your dog by howling and staring may be the way forward in improving your relationship with others, it's certainly interesting to note that when presented with the challenge, the cast had to relate to both their own sense of self, and that of a pack dynamic – which could certainly have an effect on pet owners."

One of the cast explains their experience:

"Kath led a session where we tuned into a wolf spirit.

"Now, though I have studied Dina Glouberman's IMAGEWORK and am used to working with accessing hidden intelligence via arising images and using images as mentors, it's funny how, sometimes, I discount this information as just my 'imagining.' During this session, one of the things we were asked to do was to get a sense of our role in the pack. I got a message that I was an outsider, and also received the message 'the pack can include the outsider.'

"When Kath announced the rehearsal period, I found I had a work contract during this time, and so it would seem that I could not take part in the Edinburgh run. I was surprised and deeply pleased when Kath decided to include me – I would play a role that involved me being planted in the audience, then appearing as someone who was a were-wolf character/spirit. So the downloaded message turned out to indeed be true – the pack did embrace the outsider."

There is evidence to suggest that dogs 'sense' aspects of our own personality that we keep hidden from even our nearest and dearest. Some of the dogs I have been lucky enough to work with (and not the breeds that you might consider have a sympathetic nature) have gone on to become canine counsellors, providing a bridge for depressed and bereaved people who can talk to a dog in a way they never could to a person, allowing them to explore their grief and vulnerability with a empathetic being that does not judge them.

Why are we attracted to certain breeds?

Mastitts, Molossers and wolf-like dogs have always appealed to an area of man's psyche that craves protection, manifesting at every level the physique, power and bravery that many people wish they possessed. They are testament to man's need to show and be in control of raw power (and let no owner tell you otherwise), and are the flip side to the law of attraction that 'like begets like.'

Owners of large dogs often have something to prove (and I'll come on to those who like small dogs next ...); an obvious show of strength, maybe, a misguided longing for protection, possibly. Before my postbag explodes with complaints, let me say that I am one such owner, and I know many, many others, but not one of us chose our particular shade of giant companion because we wanted a 'handbag' dog ...

I asked a very specialist group of Cane Corso owners and enthusiasts what it was that had switched them on to this breed (which is more of a suitcase dog, I guess). Disregarding its comparative rarity here in the UK (though never cited as a factor for choosing this breed, who doesn't enjoy the appeal of the rare and exotic?), I'll share with you some of the responses I received, with gratitude to the group 'The friends of Cane Corso, UK.'

Helmut and Marzena Sadowska

"Our story with the Cane Corso starts like this. It must have been somewhere in the 90s, here in Poland, and we were on a farm to buy pork, fresh from the butcher; just the meat to make sausage and something for smoking.

"On entering we get a shock because of some dogs behind the fence. Powerful? They want to break through, eyes bigger than the moon watching our steps! I have to remember this situation, so years later when my wife asks me: 'What you think about getting a dog?' I'll know what to say.

"For me it wasn't practical to get a dog. I put it out of my mind. But on my birthday I ended up getting a Cane Corso called Cinty.

"This dog changed my life. This funny girl drilled such deep anchor bolts into our hearts that I can't think of life without her.

"One dog in the home isn't enough, so my wife bought a male Cane Corso called Czedor. He was completely the opposite of Cinty, an absolute mover, nothing was sacred at our house, he damaged almost everything you can imagine; he even chewed a 220 volt power cable.

"Sadly, Cinty got Lyme Disease and died. The vet misdiagnosed her, and by the time we knew what was wrong, it was too late. I suffered for a long time over this.

"We'd both fallen for the breed and brought two more into our lives: Lux Klara and Lux Samanta. My wife returned with these two small pups and Czedor was happy to have somebody else like himself to play with.

"Then Czedor became ill. I checked his kennel one morning to find him lifeless and confused. I knew there was something wrong with him immediately. We ended up driving 500km through the night to find a vet that could do a blood test. He'd been poisoned, we think it was rat poison that did it.

"Three days passed and we didn't know if he would live; the people in that clinic were doing the absolute best they could but it was hard. After two weeks in that place we brought him home This situation showed us that there might be somebody around who didn't like our dogs. We sent the police to check out our suspicions, but still today we haven't found the evidence.

"Today, we can say that we have seen all the light and also darkness. The Cane Corso is our way, our destiny. Many years at dog shows, the fantastic moments, this life with the breed makes you crazy but also very happy. Not everyone understands this, it's more than you ever can expect from life, and something new always comes up."

Roy Gardiner

"The journey to convincing my wife that all large breed dogs are not the hounds from hell took me to Maidstone to a fund raising day. With wife, daughter and a Boston Terrier – Jack Russell cross – in tow, I led them to a small crowd of Corsos and owners.

"Whilst dealing with my wife's 'Are these dogs vicious?' line of enquiry, I found the owner handing me the lead to my first ever Corso to hold for her. From that moment when I looked into that dog's eyes for the first time, I knew I had to own a Corso. The depth and intensity of her gaze captivated me: gentle with latent power. Our home now feels complete.

"Perhaps nothing miraculous has occurred, but she has brought our family closer together. My step-daughters have been converted considering they once believed that the breed was a man-eater. My oldest step-daughter has come to handling class to work with her. And as for me, the bond I have with her is deep, she is beside me always. I move, she moves, at times it feels as if she knows what I am about to do before I do."

So, small dog owners, your turn now ...

Owners of small dogs often fall into the trap of believing they have somehow 'adopted' a baby covered in fur, whose every need they must fulfil. The fact that dogs have survived for millennia without the need for 'special prescription food' and snow bootees escapes most 'furbaby' owners.

Yet, the tiny dog, from the terrier to the lapdog, can also be the bravest of souls with the biggest personality. The tiny dog who sticks most in my mind (from a list of thousands of pint-sized heroes) was a British mongrel called Rip. Made homeless during the bombing of London in the Second World War, Rip attached himself to an air raid warden (in classic British style, I can only find him referred to as 'Mr King'). Rip soon became the prototype for search and rescue dogs everywhere – not after benefit of any training, but because, in the words of his own regiment: "We just couldn't stop him!" – how wonderfully terrier-like.

Rip saved over 100 people from being buried alive, diving into recently bombed homes with a total disregard for his own safety, and emerging covered in dust, barking and yapping, until action was taken. There are some wonderful images of him emerging from bombed buildings, with a huge grin on his face, and a dusty human clambering behind. Rip died in 1946, and is remembered with the epitaph 'Played his part in the Battle of Britain.'

Rip is far from the only mini hero, but he is my favourite because he takes us right back to that utter joy that I genuinely believe is the canine default setting. I risk the ire of my colleagues by saying that the smaller breeds of dog are infinitely more cheerful than most giants, and even slightly smarter, a notion confirmed by Dr Stanley Coren in his listing of the top 80 intelligent breeds (see chapter 5). Over half of the top 20 are miniature/small breeds, with my beloved Mastiff turning up on the last page somewhere …

Size and the sense of self

Gollum Testrom and Nancy Birch's study, *Dog adoption decision: the relationship between definition of self and dog breed choice as mediated by dog breed categorisation*, shows that, of 1129 respondents questioned both before and after adopting a new dog, a significant proportion equated the size of the animal to their perception of their own size, markedly more prevalent in adoption decisions where the male was the primary adopter:

So, does this mean that all owners of large dogs are brash men with a Napoleonic complex; or that the woman with the Chihuahua in her handbag is yearning for a baby, or missing her dollies? Of course not, reasons for being attracted to certain types of dog can be attributed to so many factors it would be impossible to cover them all here. These are extreme examples of how and why we are attracted to certain breeds, and left completely cold by others.

Owners of medium-sized, cheerful mongrels aren't always happy-go-lucky jokers without a care in the world, and the meanest individual on your street can be the owner of the most loyal dog, usually to the chagrin of those who focus more on outward appearances, whilst ignoring the social and behavioural needs of their own animal.

Dogs left to their own devices create a micro society (I am loathe to use the word 'pack,' due to its recent change of meaning to a dog/human dynamic, rather than a group of dogs/wolves) in which there is a clear hierarchy that has nothing to do with people.

Everyone has a job, a talent and a place. Sick or unstable elements are eliminated without emotion, and the group carries on with the job of survival: the best hunters hunt, the guardians guard and the trackers track. Even in feral groups, with dogs of differing size and ability, an entente cordiale is reached without grudges being held, even after a fight, and newcomers are judged on their ability to bring something to the group if they are different to the rest of the pack, an ethos we would do well to adopt.

The dog/human dynamic, wild dogs, and mating

As a species we have much to learn from the canine's loyalty and relentless good cheer, and also much to rediscover in our quest to have the best relationship we can with this animal. People say that their dog communicates with them, and I have no doubt that dogs are trying very hard to tell us something (usually "I NEED MORE EXERCISE!"), but instead of trying to understand their language, we project our own set of values onto the animal.

CLEVER DOG!

There's an important distinction between learning how a dog communicates and attempting to imitate it; you are not a dog, and nor will you ever be. Dogs are intelligent enough to understand many of the vagaries of human communication, but so thorough have their attempts to communicate with us become, they are now widely misinterpreted as emulating human emotions such as guilt ("Fido knows when he's done wrong" – he doesn't, he's just making himself small and inoffensive so you don't kill him because he can tell that you're angry), right up to the unbelievable "She doesn't fancy this boy, let's use artificial insemination."

And, of course, the dreaded AI is responsible for so many abuses of the animal world in general and the dog in particular, from the bastardisation and mutation of a breed to the point where it can no longer mate or give birth naturally, let alone perform the job it was originally created to do, to the 'we know best' argument of deciding that a certain male is the right choice for a pairing, even when the female rejects his advances.

We are doing dogs yet another massive disservice with this practice. Researching lines and performing health tests is admirable, and should be lauded in a breeding programme that aims to produce healthy animals. But when a bitch rejects a mate, she is sending a message loud and clear that, genetically, for whatever reason, this male is not the healthy/right choice to further the line. Remember; dogs are proving time and time again that they can scent locate certain cancers. The evidence is now more than anecdotal: researchers, led by György Horvath MD, PhD, from the University Hospital in Göteborg, Sweden, along with colleagues at Working Dog Clubs in Sweden and Hungary, trained dogs to distinguish different types and grades of ovarian cancer, including borderline tumours.

For argument's sake let us suppose that if we trust dogs to reliably detect cancer, drugs, weapons, money, bombs, missing people, unmarked graves, and ducks (!), they probably can reliably detect a serious genetic malfunction in a potential mate, which, if the mating went ahead and was successful, would result in the offspring carrying a malignant genome. They should be allowed to make that decision. We trust dogs to take our blind across busy streets, but not to make choices regarding their own offspring; crazy, or what?

And for us this is another lesson from our canine companions – if your gut says the guy/girl asking you on a date is no good, trust it!

Why do we choose the dogs we do?

Susan Quilliam is a highly respected relationship psychologist and author of 18 relationship books, three of which were written for Relate and The Samaritans, with whom she works closely. Susan's work helps people discover just what makes them tick, and how to enjoy better relationships with friends, colleagues and partners. Here is Susan's insight to just what motivates millions of us to enter into a relationship with a dog, and why we benefit so much from this.

Q Are there emotional and health benefits to be gained from a relationship with a dog?
A Being unconditionally loved by a dog will surely have the same benefits as a solid relationship with a human. Research suggests that as well as helping the sick recover and the old survive, pets (in this case, dogs) give the following benefits:

Emotional – raised self-esteem, reduced stress, increased confidence, feeling of being needed, sense of purpose in life, sense of control over environment.
Physical – lowered adrenalin/blood pressure/heart rate ... sense of relaxation ... boosted immune system ... release of oxytocins into the bloodstream, so creating a sense of being loved.

Q Do you believe people often have unrealistic expectations of their human companions when measured against the loyalty of a dog?
A The loyalty offered by a dog is simple and straightforward; you are the centre of their world, the one who gives them not only food but a reason for existence.

A human being – at any rate, a sane human being – will never give this sort of loyalty because, quite rightly, humans don't regard each other as the centre of each other's world. They may feel like this in the early stages of a love relationship because strong feelings of affection and loyalty are triggered by a very specific set of hormones, but after about two to four years, the obsession fades and a much more balanced and equal relationship follows. In 'normal' human contact, we are not in general dependent on each other for all our physical and emotional sustenance ... so the relationship is not so intertwined.

But that doesn't mean to say we don't want it to be. The sort of human love that many of us crave – and which pop songs and romantic myths encourage us to crave – is much more like the dependent love of a dog for a human. So when we can't get it from humans, yes, many of us are disappointed.

Q Britain has more than six million dog owners. What are the main motivations for so many people deciding to share their lives with a dog?
A Companionship is the one most often given for having a dog, along with protection and security, and fun and interaction for the children.

However, the real reasons for having a dog may be much, much deeper, because – particularly in this day and age – we so rarely receive unconditional love. The work situation is achievement-based; we have to get results in order to receive approval. And intimate relationships, friendships and family relationships are often very conditional – we only get the approval of others if we fulfil their criteria. So the canine-human relationship is unique in life, and therefore very much sought after.

Q Why has the dog, more than any other animal in history, risen to meet the expectations of mankind to such an extent that he is considered our 'best friend' rather than any of the millions of species we could have chosen from the animal kingdom?
A Dogs are uniquely built, and genetically developed from their wolf ancestry to give loyalty to the 'pack leader' – their owner, their master, their you! They follow, they adore, they understand – and, of course, historically, they protect and support the pack and its leader (you and your family).

Q Can a dog TRULY be described as someone's best friend, or is that slightly degrading to human companions?
A Of course, human companions provide a much richer variety of interactions. And we're likely to grow more personally if we interact with humans who challenge us and give us different views on the world.

But a dog's unique talent for acceptance, adoration and loyalty makes it a very good companion, if not a friend in exactly the same vein as our human cohorts.

Q What about the pain of loss? Many non-dog owners struggle to grasp how deeply affected an owner can be at the loss of a pet. Should owners feel less inhibited to admit to friends, family and colleagues just how much grief they are suffering at the loss of a dog, or should they put it into a different perspective to, for example, the loss of a good friend?
A One should always accept and admit grief after the loss of a dog; it's a real bereavement and hence we go through the classic bereavement stages of denial, shock, grief, anger, depression, and recovery. I advise people to seek counselling after the loss of a pet – I think it can be a huge blow and should be treated as such.

Telling others about it depends on who the 'others' are.

Someone who has never had pets is going to find it hard to understand just what a bereaved dog owner is going through.

However, even if someone has not had a pet, if they're sensitive and caring it's worthwhile explaining the situation and asking for support.

CLEVER DOG!

Q Why do people crave constant companionship, and what makes us lean toward certain personality types when making friends or choosing partners?

A People are pack animals, too, just as much as dogs are. We are not completely self-sufficient and historically need a tribe in order to survive. Hence, even in the modern world we are not programmed to live completely alone.

As to leaning toward certain personality types, this could fill a book in itself! In general, though, we are drawn to those who are sufficiently similar to us that we understand them, and are sufficiently different in that we complement each other and create a 'complete' whole. So an introvert may be drawn to an extrovert because he loves her bubbly personality, and she may be drawn to him because she loves his calm self-sufficiency.

In general, we are likely to choose dogs which:
• meet our practical needs; eg for protection
• parallel our values; some people may not buy breeds such as Rottweillers because they don't want to be seen as aggressive or dominant personalities themselves
• complement or reflect our emotional needs; someone in need of loyalty and affection is likely to choose a Border Collie or Labrador, for example

Lessons we can learn from dogs

So how can we make use of the many lessons the dog has to offer us, and how can we be better attuned to what he is trying to say? I went through hundreds of scientific papers during my research for this piece, some of which I have chosen to include (for all science is born of passion, regardless of the clinical outcome), some of which led me off on some amazing tangents which have provided insight and points of view beyond what I could have imagined. I make the leap that if you are reading this book, you already have a desire to learn what the dog has to teach us, and therefore will enjoy what I discovered:

Bring joy into your life: dogs do not wake up worrying about last night, nor do they obsess about today. They go outside and enjoy what is waiting for them.

Fitness makes you happy: you may not want to go whizzing through muddy puddles on a bike with your dog in the pouring rain, and *you* don't have to, because that 'inner child' that psycho-babblers the world over tell us about DOES, so let your inner child take the dog out. You might enjoy the ride!

Make yourself useful: dogs are fantastic at finding ways they can be useful to humans, from saving our lives, to lowering our heart rate and increasing our oxytocin (happy hormone) level just by being there. Find what it is you're good at doing for others – and go spread some joy!

Get listening: talk to your dog, tell him your problems and unburden yourself. Your dog won't mind and you'll feel much better for it. Then take that lesson to your loved ones. Listen and don't judge.

Give thanks: you don't have to be religious to give thanks, your dog is showing you every day that he enjoys the relationship he has with you by wanting to be where you are, and do what you want to do. Think dog, and enjoy the here and now; if you have health, food and shelter, everything else is a bonus.

Howl at the moon: this is optional. But a lot of fun ...

EPILOGUE

THE LIFE LESSONS WE CAN LEARN FROM DOGS

Throughout this book we've examined the canine condition, and I hope that it has given you a richer understanding of why so many people, myself included, find themselves drawn to dogs.

There are lots of books out there about dogs, of course, but I don't think there's anything else that has considered the dog as a role model.

I've worked with dogs for all of my adult life and, without realising it at first, have learned a lot. Every time I read a new book that teaches me something interesting, I think about how it relates to dogs. Even business books have this affect on me (dogs are great negotiators, better than people, sometimes).

Dale Carnegie said in his book *How To Win Friends and Influence People* that dogs are the finest examples of how a person should make friends. I wish Dale had taken this idea further. Dogs are indeed great at making friends; they even help us make friends with new people. Here are what I consider to be the five premier life lessons we could borrow from the dog to our advantage.

Lesson 1 Don't take life too seriously
I've never known a dog take life too seriously. I remember seeing a three-legged Greyhound hopping down the road alongside his owner, and remember feeling humbled by the realisation that this Greyhound genuinely didn't have a care in the world. His ears were slicked back, his eyes bright with alert intensity. He was looking around, sniffing around and wagging his tail. He'd lost a leg, but intuitively understood that worrying wasn't going to bring it back. If he could have talked, I think he'd have said: "I'm just going to hop as fast as my friends can run." Dogs overcome mental and physical hurdles via their simplistic approach to problem-solving.

Lesson 2 Trust your instinct
I was privileged enough to witness an explosives detection dog in action a few years ago. Her handlers had set up a test run to demonstrate her skill. At first, I thought she'd got it wrong as she was transfixed on a toilet paper holder, even though the explosives were hidden in a wall cavity. But she wouldn't budge, despite her owner's commands to carry on. It turned out the handler had been to the toilet after planting the explosive and had tainted the toilet roll holder with the scent! Dogs aren't complex enough to second-guess themselves. Nature gave them gut reactions for a reason. We have gut reactions. We should use our gut more often.

CLEVER DOG!

Lesson 3 Mean what you say
Dogs can't talk, but they also can't lie or deceive. They have absolutely nothing to gain from being misleading or dishonest, which is why you can always trust a dog. If we allow ourselves to be as genuine and committed in what we do and say as dogs are when they greet us, comfort us or attempt to please us, then I think we'd be happier.

Lesson 4 Loyalty
Leaders love dogs as much as 'ordinary' folks do. Why? A dog is loyal and offers his affections with no strings attached. He has, as one particular saying goes, more friends because he wags his tail more than his tongue. The dog is the ultimate winner of friends as he attaches no conditions to the loyalty he offers.

Lesson 5 Embrace life
The one thing I hope you take onboard from this book is this: emulate your dog's genuine and unadulterated lust for life and you'll be happier. This book can't promise you solutions to problems, or strategies for business success, but it can certainly show you how one humble species has managed to get itself a place on the board of directors in the business of life!

FURTHER READING

The Complete Dog Massage Manual – Gentle Dog Care • Robertson • 978-1-845843-22-6 • Hubble and Hattie

Dog Relax – relaxed dogs, relaxed owners • Pilguj • 978-1-845843-33-5 • Hubble and Hattie

My dog is blind – but lives life to the full! • Horsky • 978-1-845842-91-8 • Hubble and Hattie

My dog is deaf – but lives life to the full! • Wilms • 978-1-845843-81-6 • Hubble and Hattie

My dog has cruciate ligament injury – but lives life to the full! • Haüsler/Friedrich • 978-1-845843-83-0 • Hubble and Hattie

My dog has hip dysplasia – but lives life to the full! • Haüsler/Friedrich • 978-1-845843-82-3 • Hubble and Hattie

Dog Games – stimulating play to entertain your dog and you • Blenski • 978-1-845843-32-8 • Hubble and Hattie

Smellorama! – nose games for dogs • Theby • 978-1-845842-93-2 • Hubble and Hattie

WWW.HUBBLEANDHATTIE.COM

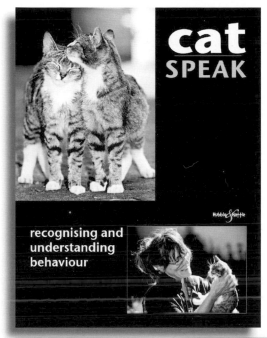

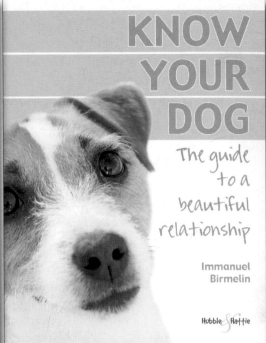
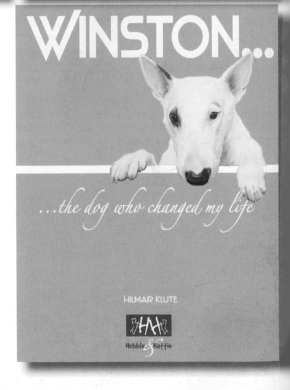